INTRODUCTION

This handbook is an easy to use and quick reference of agriculture and horticulture terminology, chemical formulas and equations, and other soil and plant related information. The book contains 1,263 terms; 43 chemical equations; plus 51 tables and figures. Every effort was used in expounding the terms, equations, and tables and figures in order for the book to be as thorough, comprehensive, and accurate as possible.

This reference is intended to be a helpful and useful companion for anyone engaged in growing plants...from the agriculture and horticulture specialist and professional, to homeowners and gardeners, growers, farm managers, and all students and educators of soil and plant sciences. The objective has been to provide a practical and accurate reference book that will be an extremely valuable resource.

For the grower and fieldman, this book is meant to be a "pickup truck companion;" ...so information pertaining to fertilizers, soil amendments, irrigation, and chemical formulas will be at hand and readily available. For the student and educator, this handbook is meant to be a quick reference, to be kept within easy reaching distance for instant access to many answers needed on a daily basis. For the home gardener, this reference is meant to help shed light on the many questions that arise when growing fruits, vegetables and ornamentals; and to make gardening more productive, fruitful, and enjoyable; and also to put to rest many myths and untruths pertaining to fertilizers, soil amendments, and other garden products.

None of the terms in **The Practical Handbook and Lexicon of Soil and Plant Sciences** are considered official, but are published in an effort to provide a foundation and understanding of terminology, soil and plant science chemistry and biology, and of agriculture and horticulture in general.

While all possible effort went into making this book completely comprehensive and 100% accurate (it has been over twenty-five years in the making), there will no doubt, be facts, formulas, and definitions that will be inadvertently omitted or overlooked. If the reader would like to contribute in rectifying any oversights or omissions, please contact the author and those additions will be included in the next edition of the book. Read, learn, and enjoy.

Brent Rouppet, Ph.D., Soil Fertility Scientist
soildoctor@fertilesoilsolutions.com
www.fertilesoilsolutions.com

The Practical Handbook and Lexicon of Soil and Plant Sciences

AAPFCO. American Association of Plant Food Control Officials. AAPFCO is an organization of fertilizer control officials from each state in the United States, and from Canada and Puerto Rico, who are actively engaged in the administration of fertilizer laws and regulations; and research workers employed by these governments who are engaged in any investigation concerning mixed fertilizers, fertilizer materials, their effect, and/or their component parts. AAPFCO strives to gain uniformity by consensus among each of these entities without compromising the needs of the consumers, protection of the environment, or fair competition among the industry.

"A" horizon, soil. (1) The A is a surface horizon in which most biological activity occurs. A Horizons may be darker in color than deeper layers and contain more organic material, or they may be lighter but contain less clay or sesquioxides. Soil organisms such as worms, nematodes, fungi, and many species of bacteria are concentrated here, often in close association with plant roots. Therefore, the A horizon may be referred to as the "biomantle." (2) Often called the "topsoil." [also see *topsoil;* and *horizon, soil*]

Abrasion. (1) The physical weathering of a rock surface by running water, glaciers, or wind loaded with fine particles. (2) The process of wearing down or rubbing away by means of friction. (3) A scraped or worn area.

Abscisic acid (ABA). Abscisic acid ($C_{15}H_{20}O_4$), also known as abscisin II and dormin, is a plant hormone that functions in many plant developmental processes, including bud dormancy. ABA promotes dormancy in perennial plants and causes rapid closure of leaf stomata when a leaf begins to wilt. ABA was originally believed to be involved in abscission (this is now known only to be the case in a small number of plants). ABA-mediated signaling also plays an import part in plant responses to environmental stress and plant pathogens. Abscisic acid is also produced in the roots in response to decreased soil water potential and other situations in which the plant may be under stress. ABA then translocates to the leaves where it rapidly alters the osmotic potential of stomatal guard cells, causing them to shrink and stomata to close. The ABA-induced stomatal closure reduces transpiration, thus preventing further water loss from the leaves in times of low water availability.

Abscission. The shedding of a body part. It most commonly refers to the process by which a plant drops one or more of its parts, such as a leaf, fruit, flower or seed. A plant will abscise a part either to discard a member that is no longer necessary, such as a leaf during autumn, or a flower following fertilization, or for the purposes of reproduction. Most deciduous plants drop their leaves by abscission before winter, while evergreen plants continuously abscise their leaves. [also see *deciduous*]

Absorption. (1) Uptake of matter (especially ions and water) by a substance. (2) The movement of ions and water into the plant roots. [Example: movement of nutrients and water into plant roots). [also see *adsorption; desorption;* and *sorption*]

> **Active absorption.** Movement of ions and water into plant roots as a result of metabolic processes by the root, frequently against an activity (or electrochemical potential) gradient.

> **Passive absorption.** Movement of ions and water into the plant root as a result of diffusion along a gradient. [also see *diffusion*]

Accelerated erosion. See *erosion, accelerated*.

Accretion. (1) The gradual increase or extension of land by natural forces acting over a long period of time, as on a beach by the washing up of sand from the sea, or on a flood plain by the accumulation of sediment deposited by a stream. (2) The addition of air particles to hydrated drops (snow, rain, sleet, etc.) by coagulation as the drops fall through the sky. [also see *aggradation*]

Accumulation (soil). The buildup or increase of one or more constituents in the soil at a given position as a result of translocation. The buildup may be a residue due to the translocation of material out of the horizon or may be due to an addition of material. Usually refers to soluble substances and clay particles.

Acicular. Needle shaped. [Example: acicular crystals; or narrow and long and pointed, as pine needles].

Acid. (1) A substance that increases the number of hydrogen ions (H^+) in a solution. (2) Chemicals that give up hydronium ions, H_3O^+ (or hydrogen ions, H^+), in a reaction to another substance. [Example: phosphoric acid gives up hydrogen ions to ammonia in the formation of ammonium phosphates] [opposite = base or alkali] [also see *acid soil (soil acidity)*]

Acid-forming fertilizer. An expression applied to any fertilizer that has a tendency to make the soil more acidic by increasing the residual acidity of a soil. Example: any fertilizer containing ammonium (NH_4^+) such as ammonium nitrate, ammonium sulfate, etc.

Fertilizers may raise or lower the pH of the growth medium. Fertilizers are rated as to their potential acidity or potential basicity. For example, 15-16-17 has a potential acidity of 215 lbs. of calcium carbonate per ton of fertilizer. This means it would take 215 lbs. of limestone to neutralize the acidic effect caused by the application of one ton of 15-16-17. On the other hand 15-0-15 has a potential basicity of 420 lbs. of calcium carbonate per ton of fertilizer. A ton of 15-0-15 would raise the pH of the growth medium as much as 420 lbs. of limestone. In each case, the larger the number the greater the potential effect the fertilizer on pH. [You can find this information on the fertilizer bag of many/most brands].

Both chemical and organic fertilizers can eventually make the soil more acid. Hydrogen is added in the form of ammonia and urea based fertilizers, and as proteins (amino acids) in organic fertilizers. Transformations of these sources of N into NO_3^- (nitrate) releases H^+ ions to create soil acidity. Therefore, fertilization with fertilizers containing ammonium or organic matter to a soil will ultimately increase the soil acidity and lower the pH. [also see *nitrification; nonacid-forming fertilizer; base fertilizer;* and *acid soil (soil acidity) residual*]

The equations for the nitrification of ammonium to nitrate:

$$2NH_4^+ + 3O_2 \xrightarrow{\text{Nitrosomonas}} 2NO_2^- + 2H_2O + 4H^+ + \text{energy}$$

$$2NO_2^- + O_2 \xrightarrow{\text{Nitrobacter}} 2NO_3^- + \text{energy}$$

> Note: For each pound of nitrogen as ammonium (or forming ammonium from urea, ammonium nitrate, and/or anhydrous ammonia) it takes approximately **1.8 pounds of pure calcium carbonate ($CaCO_3$) to neutralize the residual acidity.**

Acid rain. Rainfall or melting snow with pH less than 5.7. The acids form when sulfur dioxide and nitrogen oxides released during the combustion of fossil fuels combine with water and oxygen in the atmosphere. Strong acids such as sulfuric (H_2SO_4), nitric (HNO_3) and hydrochloric (HCl) have lowered the pH of rain and snow falling on much of northern Europe and eastern United States and Canada to between 4 and 5.

Acid rock. An igneous rock that contains more than 60% silica and free quartz (as opposed to a form of rock music focusing mainly on hallucinations and drugs).

Acid salt. Salts that release hydrogen ions when solubilized. [Example: phosphate fertilizers]

Acid soil (soil acidity). (1) Soils with pH values less than 7.0. Soil and water pH readings below 7.0 indicate an excess of hydrogen ions (H^+) over hydroxide ions (OH^-). (2) The condition of the soil (or soil solution) that contains a sufficient amount of acid substances to lower the pH below 7.0. [opposite = alkaline soil]

There are three general pools, or sources, of soil acidity: (1) active, (2) exchangeable, and (3) residual:

1. **Active acidity**: The quantity of hydrogen ions that are present in the soil water solution. The active pool of hydrogen ions is in equilibrium with the exchangeable hydrogen ions that are held on the soil's cation exchange complex. This pool most readily affects plant growth. Active acidity may be directly determined using a pH meter, such as an electron probe.
2. **Exchangeable acidity**: The amount of acid cations, aluminum and hydrogen, occupied on the CEC. When the CEC of a soil is high but has a low base saturation, the soil becomes more resistant to pH changes. As a result, it will require larger additions of lime to neutralize the acidity. The soil is then buffered against pH change.
3. **Residual (or potential) acidity**: Comprises of all bound aluminum (primarily) and hydrogen ions in soil minerals. Out of

all pools, residual acidity is least available. When the soil is limed, calcium ions (Ca^{2+}) displace aluminum ions (Al^{3+}) from the exchange sites. The Al^{3+} hydrolyzes (see equations below) and generates more H^+.

Al^{3+} (predominantly at pH values less than 4.7)

$Al^{3+} + H_2O \longrightarrow Al(OH)^{2+} + H^+$ (at pH 4.7 to 5.0)

$Al(OH)^{2+} + H_2O \longrightarrow Al(OH)_2^+ + H^+$ (at pH 5.0 to 6.5)

$Al(OH)_2^+ + H_2O \longrightarrow Al(OH)_3$ (insoluble) $+ H^+$ (at pH 6.5 to 8.5)

Acidity or Alkalinity of a Soil Measured in the Saturation Paste (pH)	
Below 4.2	Too acidic for most crops and plants
4.2 – 6.2	Suitable for some acid tolerant crops
6.2 – 7.2	Suitable for most crops
6.4	Optimum for most crops
7.2 – 7.8	Suitable for some alkaline tolerant crops
above 7.8	Calcium deficiency problems occur
above 8.3	Excessive sodium is likely a problem, but can occur at lower pH values (too alkaline for most crops)

USDA. Keys to Soil Taxonomy. 11th ed. 2010.

Acid sulfate soils. Acid sulfate soils (also called cat-clays) are naturally occurring soils, sediments or organic substrates (e.g. peat) that are formed under waterlogged conditions. These soils contain iron sulfide (FeS) minerals, predominantly as the mineral pyrite (FeS_2) or their oxidation products. In an undisturbed state below the water table, acid sulfate soils are benign. However if the soils are drained, excavated or exposed to air by a lowering of the water table, the sulfides react with oxygen to form sulfuric acid (H_2SO_4).

Acidic cations. Hydrogen (H^+) ions or cations in water that undergo hydrolysis to form an acidic solution, as do aluminum (Al^{3+}) and iron (Fe^{3+}). [also see *hydrolysis*]

Acidifier, soil. A material or mixture used, especially in arid and semi-arid areas, to neutralized soil alkalinity. Sulfuric acid (H_2SO_4), phosphoric acid (H_3PO_4), liquid sulfur dioxide (SO_2), aluminum sulfate [$Al_2(SO_4)_3$], elemental sulfur (S), and ammonium polysulfide [$(NH_4)_2S$] are all soil acidifiers.

Acidity, active. See *acid soil (soil acidity) active acidity*.

Acidity and basicity of fertilizers. Acidic residue of any fertilizer in the soil is measured in terms of the calcium carbonate required to neutralize it. The basic residue of any fertilizer is measured in terms of calcium carbonate equivalent.

Principal Acid Forming Nutrients (equivalent acidities)		Principal Base Forming Nutrients (equivalent basicities)	
Relative Amounts (pounds)		Relative Amounts (pounds)	
sulfur	63	calcium	50
chlorine	28	magnesium	82
phosphoric acid (H_3PO_4)	14	potassium oxide (K_2O)	22
nitrogen	36	sodium	43

Brady and Weil. Elements of the Nature and Properties of Soils. 3rd ed. 2008.

Acidulated bone. A fertilizer made from ground bone or bone meal that has been treated with sulfuric acid.

Acidulated fish tankage. A fertilizer that is derived from rendered fish or fish scrap treated with sulfuric acid.

Acidulation. The process of treating a material with an acid. The most common acidulation process is treatment of phosphate rock with an acid or mixture of acids to increase phosphorus availability.

Acre. An area of 43,560 square feet, or 0.405 hectare.

Acre-foot. The volume of water required to cover 1 acre of land (43,560 square feet) to a depth of 1 foot. Equal to 325,850.58 U.S. gallons or 1,233.5 cubic meters.

Acre-furrow slice. Usually the top 6 inches of soil. An acre-furrow slice weighs two million (2×10^6) pounds if the soil bulk density is 1.47 grams per cubic centimeter (g cm^{-3}) [or 91.83 pounds per cubic foot (lbs ft^{-3})].

Actinomycete. A non-taxonomic term applied to a group of filamentous bacteria with characteristics intermediate between simple bacteria and the true fungi. Actinomycetes aid in the decomposition of organic matter, especially cellulose and other resistant organic molecules. Nearly 500 antibiotics have been isolated from actinomycetes, the most common of which are Streptomycin, Terramycin, and Neomycin, and Tetracycline. [also see *microorganism, soil (microbe);* and *geosmin*]

Activated sewage. A fertilizer made from sewage freed from grit and coarse solids and aerated after being inoculated with microorganisms. The resulting flocculated organic matter is withdrawn from the tanks, filtered with or without the aid of coagulants, dried, ground and screened.

Active Acidity. See *acid soil (soil acidity).*

Adenosine triphosphate (ATP). (1) A common form in which energy is stored in living systems. ATP, with the chemical formula ($C_{10}H_8N_4O_2NH_2$), consists of a nucleotide (with ribose sugar) with three phosphate groups. (2) The major source of usable chemical energy of metabolism in the cell.

Adhesion. (1) Molecular attraction that holds the surfaces of two dissimilar substances in contact. (2) The ability of molecules of one substance to adhere to a different substance. [Example: water and soil particles]. [also see *cohesion*]

Adsorption. (1) The bonding, usually temporary, of ions or compounds to the surfaces of a solid, such as calcium ions (Ca^{2+}) held on the surface of a clay crystal. (2) Process by which atoms, molecules, or ions are taken up and retained on the surfaces of solids by chemical or physical binding. Soil colloids adsorb large amounts of ions and water. [also see *sorption; absorption;* and *desorption*]

Adsorption complex. The various substances in the soil capable of adsorption, mainly clay and humus.

Adsorption, specific. The strong adsorption of ions or molecules on a surface. Specifically, adsorbed materials are not readily removed by ion exchange.

Adventitious. Referring to a structure arising from an unusual place, such as buds at other places than leaf axils, or roots growing from stems or leaves. [Example: adventitious roots; roots that develop from the stem following the death of the primary root. Branches from the adventitious roots form a fibrous root system in which all roots are about the same size; occur in monocots]

Aeolian (obsolete). See *eolian*.

Aeration, soil. (1) The condition, and sum of all processes affecting, soil pore-space gaseous composition, particularly with respect to the amount and availability of oxygen for use by soil biota and/or soil chemical oxidation reactions. (2) The process by which air in the soil is replaced by air from the atmosphere. (3) To allow or promote exchange of soil gases with atmospheric gases. Poorly aerated soils usually contain much higher percentages of carbon dioxide (CO_2) and correspondingly lower percentage of molecular oxygen (O_2) than the atmosphere above the soil.

Aerobic. (1) Containing air, and specifically referring to oxygen content in the soil rhizosphere. (2) Having molecular oxygen (O_2) as a continuous supply or part of the environment. (3) Growing only in the presence of molecular oxygen, such as aerobic organisms. (4) Occurring only in the presence of molecular oxygen, especially with chemical or biochemical processes such as aerobic decomposition. [opposite = anaerobic] [also see *rhizosphere*]

Aerobic composting. Composting environments characterized by bacteria active in the presence of oxygen (aerobes). Aerobic composting generates more heat and is a faster process than anaerobic composting. Aerobic temperatures may reach over 140 °F-- high enough to destroy pathogens, weed seeds, and fly ova. Aerobic composting creates no excessive unpleasant odors; and the most rapid composting process occurs with enclosed aerobic systems. [also see *anaerobic composting*]

Aerobic respiration. Where organisms utilize oxygen to break down components, derive energy, and generate needed biomolecules. Carbohydrates are cycled into water and carbon dioxide.

Agglomeration. A processing step in the granulation of fertilizers assembling small particles into larger granules.

Aggradation. The building-up of the Earth's surface by deposition of sediment; specifically, the accumulation of material by any process in order to establish or maintain uniformity of grade or slope. Aggradation occurs in areas in which the supply of sediment is greater than the amount of material that the system is able to transport. Typical aggradational environments include lowland alluvial rivers, river deltas, and alluvial fans. [also see *accretion*]

Aggregate. (1) A group or discrete cluster of primary soil particles (sand, silt, clay) that cohere to each other more strongly than to other surrounding particles. (2) A unit of soil structure generally less than 10 millimeters in diameter and formed by natural forces and substances derived from root exudates and microbial products which cement smaller particles into larger units.

Aggregation, soil. The cementing or binding together of several primary soil particles (sand, silt, clay) into secondary units, aggregates, or granules, usually by natural forces and substances derived from root exudates, microbial activity (e.g., by humus). [also see *humus*; *flocculation*]

Agriculture. (1) The series of processes whereby a given area of land is artificially induced to yield food for more animals and people than it would naturally support. (2) An activity of man carried out primarily to produce food and fiber and fuel, as well as many other materials by the deliberate and controlled use of mainly terrestrial plants and animals.

Agricultural lime. See *lime, agricultural*.

Agricultural mineral. See *mineral, agricultural*.

Agricultural slag. See *slag, agricultural*.

Agricultural waste. See *waste, agricultural*.

Agroecosystem. Land used for crops, pasture, and livestock; the adjacent uncultivated land that supports other vegetation and wildlife; and the associated atmosphere, the underlying soils, groundwater, and drainage networks.

Agronomy. (1) The theory and practice of crop production and soil management (as defined by the Soil Science Society of America). (2) A specialization of agriculture concerned with field crop production and soil management. The scientific management of land.

Air (porosity), soil. (1) The soil atmosphere; the gaseous phase of the soil, being that volume not occupied by solid or liquid. (2) The proportion of the bulk volume of soil that is filled with air at any given time or under a given condition, such as a specified moisture potential; usually the large (or macro-) pores.

Algae (singular alga). Soil algae, a major division of the plant kingdom, are simple, rootless, microscopic, unicellular or multicellular plants containing chlorophyll that carry on photosynthesis. The main groups are green algae, yellow-green algae, and diatoms.

Some algae are aquatic, or occur in damp situations and include most seaweeds. Some algae live and grow in sunlit waters in proportion to the amount of available nutrients. Algae can affect water quality adversely by lowering the dissolved oxygen in the water. They are also food for fish and small aquatic animals.

Blue-green algae (obsolete) [Greek *kyanos* = blue]. What were once called blue-green algae have been reclassified into the Monera kingdom as cyanobacteria. They are one of the groups of photosynthetic microorganisms. [also see *cyanobacteria*]

Brown algae. Multicellular protists placed in the division Phaeophyceae (or *Phaeophyta*), includes kelp.

Green algae. The common name for the *Chlorophyceae* (or *Chlorophyta*). A large class of algae which have chlorophylls and carotinoids (fat-soluble pigments) similar to those of higher plants and appear green, store food as starch and have cellulose cell walls.

Yellow-green algae. Algae in the division *Chrysophyta* with its chlorophyll masked by yellow pigment.

Algal blooms. Sudden spurts of algal growth, which can affect water quality adversely and indicate potentially hazardous changes in local water chemistry. [also see *eutrophication*]

Alkali. Also called "base." (1) A basic ionic salt (or compound) of hydroxide ions (OH⁻) with one of the basic cations sodium (Na⁺), potassium (K⁺), ammonium (NH₄⁺), etc. (2) Chemicals that take up hydrogen ions (H⁺). (3) A basic, ionic salt of an alkali metal that dissolves in water. A solution of a alkali, or base, has a pH greater than 7. [Examples: sodium hydroxide or potassium hydroxide] [opposite = *acid*] [also see *base*; *basic cations*; *black alkali*; and *alkali (or sodic) soil*]

Alkali metal. The Group 1 elements, lithium (Li), sodium (Na), potassium (K), rubidium (Rb), cesium (Cs), and francium (Fr). The alkali metals are all highly reactive and are never found in elemental forms in nature. All of the alkali metals are notable for their vigorous reactions with cold water to form strongly alkaline hydroxide solutions. Their chemical reactions with water are as follows:

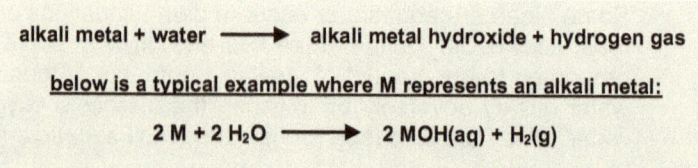

alkali metal + water ⟶ alkali metal hydroxide + hydrogen gas

below is a typical example where M represents an alkali metal:

$2M + 2H_2O \longrightarrow 2MOH(aq) + H_2(g)$

Alkali (or sodic) soil. (1) clay soils where the soil's pH is usually high, often above 9.0. A soil pH above 8.4 typically indicates that a sodium problem exists. These soils also have poor soil structure and a low infiltration capacity. Often they have a hard calcareous layer at 0.5 to 1 meter depth. Alkali soils owe their unfavorable physico-chemical properties mainly to the dominating presence of sodium carbonate (Na_2CO_3) and/or sodium bicarbonate ($NaHCO_3$) which causes the soil to swell (deflocculate). (2) Soils that contain sufficient sodium to interfere with the growth of most crop plants. Sometimes alkali soils are black due to the mixture of humic substances. "Black alkali" refers to a sodic soil condition where organic matter has spread and is present as a dusty material on the soil surface. [also see *deflocculation, soil*]

The natural cause of sodic soil is the presence of soil minerals producing sodium carbonate upon weathering. The man-made cause is the application of irrigation water (surface or ground water) containing a relatively high proportion of sodium bicarbonates. (also see *black alkali soil*; *saline-sodic soil;* and *sodic* soil)

Alkaline (or basic) soil. (1) Any soil with pH values greater than 7.0 indicating an excess of hydroxide ions (OH^-) over hydrogen ions (H^+). (2) The condition of water or soil that contains a sufficient amount of alkali substances to raise the pH above 7.0. [opposite = acid soil] [also see *sodic soil*]

Alkalinity. (1) The quality or state of being alkaline. (2) The range of the pH scale above pH 7.0. An alkali or base, when dissolved, results in alkalinity or an alkaline pH (above 7.0).

Acidity or Alkalinity of a Soil Measured in the Saturation Paste (pH)	
Below 4.2	Too acid for most crops and plants
4.2 – 6.2	Suitable for some acid tolerant crops
6.2 – 7.2	Suitable for most crops
6.4	Optimum for most crops
7.2 – 7.8	Suitable for some alkaline tolerant crops
above 7.8	Calcium deficiency problems occur
above 8.3	Excessive sodium is likely a problem, but can also occur at lower pH values (too alkaline for most crops)

Brady and Weil. Elements of the Nature and Properties of Soils. 3rd ed. 2008.

Alkaliphile. Alkaliphiles are microbes, classified as "extremophiles," that thrive in alkaline environments with a pH of 9 to 11 such as playa lakes and carbonate-rich soils. To survive, alkaliphiles maintain a relatively low alkaline level of about 8 pH inside their cells by constantly pumping hydrogen ions (H^+) in the form of hydronium ions (H_3O^+) across their cell membranes into their cytoplasm.

Allelopathy. The action of some substance produced by or in one plant species that reduces the growth of another plant species. [Example: alkaloids and other chemicals in Eucalyptus leaves retarding the growth of neighboring grass species]. [compare to *antinutrients*]

Alluvial. Pertaining to material or processes associated with transportation and/or subaerial deposition by concentrated running water. [also see *subaerial*]

Alluvium (alluvial soil). Eroded soil sediments deposited from flowing water. A general term for all detrital material deposited or in transit by streams, including gravel, sand, silt, clay, and all variation and mixtures of these. Alluvium is generally unconsolidated. [also see *detritus*]

Aluminum (Al). A widely distributed element commonly found combined in silicates in various clays and rocks; comprising about 8% of most temperate agricultural mineral soils. While aluminum may be essential to the growth of some plants, the amount required, if any, is very small. Aluminum supply in all soils is abundant. Some acid soils contain sufficient exchangeable aluminum to be toxic to plants.

When soils are too acidic (especially when pH values are equal to or less than 5.4), aluminum that is locked up in clay minerals dissolves into the soil as toxic ions, making it hard for most plants to grow. Aluminum toxicity in acidic soils limits crop production in as much as half the world's arable land, mostly in developing countries in Africa, Asia and South America. Management of soil pH is the key factor in avoiding aluminum toxicities. Aluminum toxicity may be corrected by liming. [Note: <u>Aluminum toxicity symptoms in agricultural plants may manifest as phosphorus deficiency</u>]. [also see *acid soil (soil acidity) 3. residual (or potential) acidity*]

Amendment, packaged soil. Any substance distributed for the purpose of promoting plant growth or improving the quality of crops by conditioning soil solely through physical means. This category includes all of the following:

- hay
- straw
- peat moss
- leaf mold
- sand
- wood products
- any product intended for use as a potting medium planting mix, or soilless growing media
- manures sold without guarantees for plant nutrients
- any other substance or product which is intended for use solely because o its physical properties
[also see *amendment, soil*]

Amendment, soil. Any material such as lime, anhydrite, gypsum, compost or synthetic conditioner that is applied to the soil surface or incorporated into the soil, in <u>order to make the soil more productive and enhance plant growth</u>. Soil amendments may contain important fertilizer elements but the term commonly refers to added materials other than those used primarily as fertilizers. Example: an important use of a soil amendment would be the application of anhydrite or gypsum into sodic soils to displace sodium from the exchange complex and improve soil permeability. [also see *conditioners, soil*]

Amino acid. An organic compound containing one amino group (-NH_2) and at least one carboxyl group (-COOH). In addition, some amino acids (cystine, cysteine, and methionine) contain sulfur. Amino acids are critical to life and have many functions in metabolism. One particularly important function is to combine and serve as the building blocks of proteins (linear chains of amino acids). Therefore, amino acids are a fundamental constituent of all living matter. Amino acids are synthesized by autotrophic organisms, principally green plants. [also see *autotrophic organism;* and *protein*]

Amino group. An -NH_2 group attached to a carbon skeleton as in the amines and amino acids.

Ammonia (NH_3). (1) Agricultural anhydrous ammonia fertilizer, an economical source of nitrogen widely used for direct application to soil and in irrigation waters. NH_3 is a strongly alkaline chemical compound (pH = 10.6 – 11.6 when dissolved in water) composed of 82.25% nitrogen (the highest amount of any N fertilizer) and 17.75% hydrogen, and occurs in the form of a compressed gas. Because it is stored as a pressurized liquid, special storage, handling, and application equipment is required. In some respects NH_3 behaves like water, since they both have solid, liquid, and gaseous states. The great affinity of anhydrous NH_3 for water is apparent from its solubility. As a result, NH_3 is rapidly absorbed by water in human tissue and is a serious hazard; safety goggles, rubber gloves, and an NH_3 gas mask are required safety equipment. (2) Ammonia is one of the most important inorganic nitrogen compounds in atmospheric water droplets. Ammonia reacts with strong acids and is one of the only known basic gas phase atmospheric components. The major sources of ammonia are decaying natural organic matter, livestock wastes, fertilizers, and industrial activity.

Ammonia fixation. This chemical fixation of ammonia is the entrapment, or other strong binding, of ammonium (NH_4^+) ions by organic matter and certain silicate clays. This type of fixation (by which K^+ ions are similarly bound) is not to be confused with the very beneficial biological fixation of atmospheric nitrogen gas into compounds usable by plants. In agricultural soils, 5% to 20% of the total nitrogen is found as fixed ammonium ions.

Ammonia liquor (NH_4OH) (20-0-0). Liquor ammonia (also called aqueous ammonia, or aqua ammonia) is produced by dissolving anhydrous ammonia in water. It is a colorless liquid with a very pungent odor. Aqua ammonia is an excellent acid neutralizer whose pH varies with concentration [Example: a 29% solution of NH_4OH has a pH = 13.9]. [also see *aqua ammonia*]

Ammonia volatilization. The chemical process whereby nitrogen is lost to the atmosphere as gaseous ammonia (NH_3). Volatilization greatly increases with soil pH values of 8 and above. [also see *ammonification;* and *urease inhibitor*]

Ammoniated superphosphate ($Ca(NH_4H_2PO_4)_2$) (5-40-0). (1) A fertilizer formed when superphosphate is treated/reacted with ammonia or with solutions which contain ammonia and other compounds of nitrogen. (2) A fertilizer containing 5 parts of ammonia to 100 parts of superphosphate. The guaranteed percentages of nitrogen and of available phosphate is stated as part of the name. [see *superphosphate*]

Ammoniation. A process wherein anhydrous ammonia, aqua ammonia or a solution containing ammonia and other forms of nitrogen, is used to treat superphosphate to form ammoniated superphosphate, or to treat a mixture of fertilizer ingredients (including superphosphate) in the manufacture of a multiple-nutrient fertilizer.

Ammonification. The biochemical process whereby inorganic ammonia (NH_3) and ammonium (NH_4^+) is produced by the decomposition of nitrogen containing organic compounds (amino acids, proteins, etc.). The released inorganic nitrogen is not necessarily lost from the soil system. [also see *ammonia volatilization; nitrification;* and *urease inhibitor*]

The ammonification of organic compounds/matter is shown with the chemical reactions below:

$$\text{organic compounds} + O_2 \xrightarrow{\text{enzymes (from bacteria)}} CO_2 + H_2O + \text{energy} + \text{nutrients (e.g., } NH_4^+ \text{ [ammonium])} + \text{humus}$$

$$2NH_4^+ + 3O_2 \xrightarrow{\text{Nitrosomonas}} 2NO_2^- \text{(nitrite)} + 2H_2O + 4H^+ + \text{energy}$$

$$2NO_2^- + O_2 \xrightarrow{\text{Nitrobacter}} 2NO_3^- \text{(nitrate)} + \text{energy}$$

Ammonium (NH_4^+). The univalent chemical ion derived from ammonia, whose compounds chemically resemble the alkali metals. The ammonium ion is mildly acidic. When ammonia is dissolved in water, a tiny amount of it converts (protonation) to ammonium ions as shown below:

$$NH_3(aq) + H_2O \rightleftharpoons NH_4^+(aq) + OH^-(aq)$$

The degree to which ammonia forms the ammonium ion depends on the pH of the solution. If the pH is low, the equilibrium shifts to the right: more ammonia molecules are converted into ammonium ions. If the pH is high (the concentration of hydrogen ions is low), the equilibrium shifts to the left: the hydroxide ion abstracts a proton from the ammonium ion, generating ammonia.

Ammonium hydroxide (NH_4OH) (20-0-0). See *aqua ammonia*.

Ammonium nitrate (NH_4NO_3) (33-0-0) . A fertilizer that is chiefly the ammonium salt of nitric acid. Ammonium nitrate should contain not less than thirty-three percent nitrogen, one-half of which is in the ammonium form and one-half in the nitrate form. It is also the main component of ANFO (ammonium nitrate/fuel oil), a very popular explosive.

Ammonium phosphate (($NH_4)_3PO_4$) (16-20-0). A fertilizer obtained when phosphoric acid is treated with ammonia (anhydrous or aqueous), and consists principally of monoammonium phosphate and diammonium phosphate or a mixture of these two salts. The guaranteed percentage of nitrogen and of available phosphate should be stated as part of the name.

Ammonium phosphate nitrate (($NH_4)_3PO_4$ + NH_4NO_3). A mixture of ammonium phosphate and ammonium nitrate. Produced by ammoniating the solution separated from phosphate rock that has been acidulated with an excess of nitric acid. Average composition is about 27% N and 20% available P_2O_5. Example: 28-14-0. [also see *ammoniation;* and *acidulation*]

Ammonium phosphate sulfate (($NH_4)_3H_2PO_4SO_4$). A fertilizer obtained when a mixture of phosphoric acid and sulfuric acid is neutralized with ammonia. It consists principally of a mixture of ammonium phosphate and ammonium sulfate. The guaranteed percentages of nitrogen and of available phosphate should be stated as a part of the name. The 13-39-0 grade contains about 20% ammonium sulfate.

Ammonium sulfate (($NH_4)_2SO_4$] (21-0-0). A solid fertilizer that is chiefly the ammonium salt of sulfuric acid, manufactured by reacting ammonia with sulfuric acid. Ammonium sulfate contains 21% nitrogen as ammonium cations, and 24% sulfur as sulfate anions. Ammonium sulfate is one of agriculture's oldest solid forms of fertilizer.

Ammonium sulfate nitrate (NH_4NO_3 + ($NH_4)_2SO_4$). A fertilizer that is a double salt of ammonium sulfate and ammonium nitrate which are present in equal molecular proportions. It should contain not less than twenty-six percent nitrogen, one-fourth of which is in the nitrate form and three-fourths in the ammonium form.

Ammonium thiosulfate (($NH_4)_2S_2O_3$). A commercial fertilizer composed principally of $(NH_4)_2S_2O_3$. The guaranteed percentages of nitrogen and sulfur should be stated as part of the name. Ammonium thiosulfate contains 43.1% sulfur and 18.9% nitrogen.

Amoeba (plural = amoebae). Protozoa that can alter their cell shape, usually by the extrusion of one or more pseudopodia (a protrusion of cytoplasm serving for locomotion).

Amorphous. Meaning without form. An "amorphous solid" is a solid which has a disordered atomic structure. In contrast, solids that exhibit long-range order in the atomic positions are called crystallines or morphous structures.

Anabolism. (1) Metabolic processes involved in the synthesis of cell constituents from simpler molecules. (2) The constructive part of metabolism; the total chemical reactions involved in biosynthesis. (3) The synthesis of complex molecules and new protoplasm. An anabolic process usually requires energy. [also see *catabolism;* and *metabolism*]

Anaerobic. (1) A deficiency or lack of air, and specifically referring to oxygen content in the soil rhizosphere. (2) The absence of molecular oxygen as a part of the environment. (3) Growing only in the absence of molecular oxygen, such as anaerobic bacteria. Strict, or obligate, anaerobes cannot survive in the presence of oxygen. [opposite = aerobic] [also see *rhizosphere*]

Anaerobic composting. Composting environments where the microorganisms obtain oxygen from the waste itself since they are active in the absence of oxygen (anaerobes). Peak temperatures may reach 100 °F to 130 °F, and digestion requires more time than aerobic composting. Foul odors (especially hydrogen sulfide [rotten egg odor], cadavarine, and putrescine) are created with anaerobic composting since pathogens may survive. [also see *aerobic composting*]

Anaerobic decomposition. Reduction of the net energy level and change in chemical composition of organic matter caused by microorganisms in an oxygen-free environment.

Anaerobic organism. (1) Refers to organisms that are not dependent on oxygen for respiration. (2) An organism that lives in an environment without molecular oxygen.

Anaerobic respiration. Living or acting in the absence of oxygen. Cellular respiration in the absence of oxygen.

Analysis (fertilizer). see *guaranteed analysis (fertilizer)*.

Angle of repose. (1) The inclination of a plane at which a body placed on the plane would remain at rest, or if in motion would roll or slide down with uniform velocity; the angle at which the various kinds of earth will stand when abandoned to themselves. (2) The angle between the horizontal and the plane of contact between two bodies when the upper body is just about to slide over the lower. Also known as angle of friction. In studies of sedimentation, the angle at which granular material comes to rest. The angle of repose of sand, for example, is between 30° and 35°.

Anhydride. An oxide that will react with water to form the corresponding acid or base. P_2O_5 is an anhydride of H_3PO_4; CaO is the anhydride of $Ca(OH)_2$.

Anhydrite (anhydrous calcium sulfate) ($CaSO_4$). A chemical sedimentary rock with a solubility of 0.205 grams per 100 grams water (2003-2004 CRC Handbook of Chemistry and Physics), and typically, a 100% gypsum equivalent. Same uses in agriculture and horticulture as gypsum ($CaSO_4 \cdot 2H_2O$). A typical analysis of high quality anhydrite contains 25.2% calcium and 20.0% sulfur. As a soil amendment, often called the "miracle amendment" since uses and benefits include: (a) a fertilizer source for calcium and sulfur, (b) improving soil structure, (c) improving, amending and reclaiming soils high in destructive sodium, (d) replacing harmful salts such as sodium, (e) improving water use efficiency, (f) reducing runoff and erosion, (g) preventing soil crusting, (h) improving compacted soils, and (i) preventing water logging of soils.

The addition of solution-grade anhydrite to irrigation water will help replace any calcium precipitated as lime due to high concentrations of bicarbonate. Anhydrite is one of the few materials added to soils that qualifies as <u>a fertilizer, a soil amendment, and a soil conditioner</u>. [also see *bicarbonate; gypsum; hemihydrate; amendment, soil;* and *conditioner, soil*]

Solubility of Anhydrite
0.205 grams/100 grams water = 0.0171 pounds per gallon
(5,575 pounds per acre foot water)

1 milliequivalent anhydrite per liter water ≈ 234 pound per acre-foot

Therefore: The maximum solubility of anhydrite in one acre-foot of water is:
23.8 milliequivalents per liter (or 5,575 pounds per acre foot water

---

CRC Handbook of Chemistry and Physics. 91st ed. 2010.

Anhydrous ammonia. See *ammonia (NH_3)*.

Anhydrous calcium sulfate. See *anhydrite (anhydrous calcium sulfate) ($CaSO_4$)*.

Anion. Ions in solution with one or more atoms having a negative electrical charge. Nonmetals usually form anions. [Examples: nitrate (NO_3^-), sulfate (SO_4^{2-}) and chloride (Cl^-)]. [also see *cation*]

Plant Nutrients and Other Important Anions in Soils and Irrigation Water		
Element Name	Chemical Symbol	Ionic Forms in Soils and Water
Boron	B	H_3BO_3, HB_4O_7, BO_3^-, $B_4O_7^-$
Carbon	C	CO_3^{2-} (carbonate) HCO_3^- (bicarbonate)
Chlorine	Cl	Cl^-
Molybdenum	Mo	MoO_4^{2-}, $HMoO_4^-$
Nitrogen	N	NO_3^- (nitrate)
Oxygen	O	O^{2-} OH^- (hydroxide)
Phosphorus	P	$H_2PO_4^-$, HPO_4^{2-} (orthophosphates)
Selenium	Se	SeO_4^{2-}
Sulfur	S	SO_4^{2-} (sulfate)

Compiled From Various Sources

Anion exchange capacity. The sum total of exchangeable anions that a soil can adsorb. Usually expressed as milliequivalents per 100 grams of dry soil or centimoles per kilogram of soil.

Annelid. A red-blooded worm, the annelids (also called "ringed worms"), formally called Annelida, are a large phylum of segmented worms, with over 17,000 modern species including ragworms, earthworms and leeches. They are found in marine environments from tidal zones to hydrothermal vents, in freshwater, and in moist terrestrial environments.

Annual plant. (1) A plant in which the life cycle is completed in a single growing season. (2) Plants that grow and reproduce sexually during one year. [also see *biennial;* and *perennial*]

Anoxic. Literally "without oxygen." An adjective describing a microbial habitat devoid of oxygen. The lack of oxygen such as the inadequate oxygenation of the blood (anoxia). In aquatic environmental chemistry it refers to water that has become oxygen poor due to the bacterial decay of organic matter.

Anthocyanins. Water-soluble vacuolar pigments, with the molecular formula $C_{15}H_{11}O^+$, that may appear red, purple, or blue according to soil pH. They belong to a parent class of molecules called flavonoids synthesized via the phenylpropanoid pathway; they are odorless and nearly flavorless, contributing to taste as a moderately astringent sensation. Anthocyanins occur in all tissues of higher plants, including leaves, stems, roots, flowers, and fruits.

In flowers, bright reds and purples are adaptive for attracting pollinators. In fruits, the colorful skins also attract the attention of animals, which may eat the fruits and disperse the seeds. In photosynthetic tissues (such as leaves and sometimes stems), anthocyanins have been shown to act as a "sunscreen," protecting cells from high-light damage by absorbing blue-green and UV light, thereby protecting the tissues from photoinhibition, or high-light stress.

Anthropogenic. Refers to something originating from humans and the impact of human activities on nature.

Antibiotics. Substances produced by some microorganisms (e.g., actinomycetes), plants, and vertebrates that kill or inhibit the growth of bacteria. Nearly 500 antibiotics have been isolated from actinomycetes, the most common of which are Streptomycin, Terramycin, and Neomycin, and Tetracycline.

Antinutrients. Chemicals produced by plants as a defense mechanism; inhibit the action of digestive enzymes in insects that attack and attempt to eat the plants. Phytic acid, polyphenols and oxalic acid are examples of antinutrients. [compare to *allelopathy*]

AOAC International. Association of Official Agricultural Chemists. An international, non-profit, scientific association with headquarters in Gaithersburg, Maryland. AOAC publishes standardized chemical analysis methods designed to increase confidence in results of chemical and microbiological analyses. Government agencies and civil organizations often require that laboratories use official AOAC methods.

Apatite (or rock phosphate). A group of phosphate minerals having the formula $Ca_{10}(X_2)(PO_4)_6$, where X is usually fluorine, chlorine, or the hydroxyl group, either singly or together. Fluorapatite is widely distributed as the crystalline mineral and as amorphous phosphate rock, both forms of which are important fertilizer materials.

Apical meristem. Embryonic tissue at the tip of a shoot or root that is responsible for increasing the plant's length.

Apparent specific gravity. The weight of a fertilizer divided by the weight of an equal volume of water at 40 °F, expressed in grams per cubic centimeter or pounds per cubic foot. Apparent specific gravity determine the rate at which an applicator setting will apply a given fertilizer to a soil.

Aqua ammonia [ammonium hydroxide (NH$_4$OH)] (20-0-0). Also called ammonia liquor or liquor ammonia. A liquid fertilizer formed by dissolving anhydrous ammonia in water. Commercial grades of aqua ammonia usually contain 20 percent-25 percent nitrogen. Aqua ammonia is an excellent acid neutralizer whose pH varies with concentration [Example: a 29% solution of NH$_4$OH has a pH = 13.9].

Aquaculture. Farming of plants and animals that live in water, such as fish, shellfish, and algae.

Aqueous. A solution in which the solvent is water. It is usually shown in chemical equations by appending (aq) to the relevant formula. The word aqueous means pertaining to, related to, similar to, or dissolved in water.

Aqueous solubility. The maximum concentration of a chemical that will dissolve in pure water at a reference temperature. [Example: the aqueous solubility of anhydrite (CaSO$_4$) is 0.205 grams/100 grams water]

Aqueous solution. Any solution in which water is the solvent. [Examples: cola, saltwater, and rain (H$_2$CO$_3$)]

Aquifer. (1) A geologic formation that is water bearing. 2) A geological formation or structure that stores and/or transmits water, such as to wells and springs. Use of the term is usually restricted to those water-bearing formations capable of yielding water in sufficient quantity to constitute a usable supply for people's uses.

(confined aquifer). Soil or rock below the land surface that is saturated with water. There are layers of impermeable material

both above and below it and it is under pressure so that when the aquifer is penetrated by a well, the water will rise above the top of the aquifer.
(unconfined aquifer). An aquifer whose upper water surface (water table) is at atmospheric pressure, and thus is able to rise and fall.

Arable land. Land suitable for the production of cultivated crops. According to an FAO (Food and Agriculture Organization of the United Nations) report, the global land area without major soil fertility constraints is about 31.8 million square kilometers, and total potential arable land is about 41.4 million square kilometers.

Arid (climate). A term applied to regions or climates that lack sufficient moisture for crop production without irrigation. In cool regions annual precipitation is usually less than 25 centimeters (approximately 10 inches). It may be as high as 50 centimeters (approximately 20 inches) in tropical regions. Natural vegetation is desert shrubs. [also see *semiarid*]

Aridity. The condition of receiving sparse rainfall; associated with cooler climates because cool air can hold less water vapor than warm air. Many deserts occur in relatively warm climates, however, because of local or global influences that block rainfall.

Artesian water. Ground water that is under pressure when tapped by a well and is able to rise above the level at which it is first encountered. It may or may not flow out at ground level. The pressure in such an aquifer commonly is called artesian pressure, and the formation containing artesian water is an artesian aquifer or confined aquifer. [also see *aquifer*]

Arthropod. A member of the phylum arthropoda, which is the largest in the animal kingdom. Arthropods are invertebrate animals with jointed legs; and include insects, such as ants and springtails; arachnids, such as spiders and mites; myriapods, such as centipedes and millipedes; and crustaceans, such as sow bugs.

Artificial recharge. A process where water is put back into ground-water storage from surface-water supplies such as irrigation, or induced infiltration from streams or wells.

Asbestos. Essentially a trade name given to naturally occurring fibrous silicate minerals used in commercial products. The most common minerals used to produce asbestos are chrysotile, crocidolite, actinolite, tremolite, amosite, and anthophyllite. Each of these minerals has its own formula. Asbestos itself is not a mineral so it does not have a chemical formula.

The primary use of asbestos was as a fireproofing material. After widespread use of asbestos, over a 15 to 25 year time span, a pattern of illness gradually began to occur in asbestos workers. Three diseases linked to asbestos exposure are asbestosis, a fibrous scarring of the lungs, lung cancer, and mesothelioma, a cancer of the lining of the chest or abdominal cavity.

Ash. The mineral content of a product remaining after complete combustion. The grayish-white to black powdery residue left when something is burned. Wood ash contains calcium carbonate as its major component, representing 25 or even 45 percent, less than 10 percent is potash, and less than 1 percent phosphate. There are trace elements of iron, manganese, zinc, copper and some heavy metals. However these numbers vary as combustion temperature and original organic product are important variables in determining wood ash composition.

Ash (volcanic). Unconsolidated, pyroclastic material less than 2 millimeters in all dimensions. Commonly called "volcanic ash." Sometimes the particles are shot high into the atmosphere and carried long distances by the wind.

Aspect. The direction toward which a slope faces with respect to the compass or to the rays of the sun; also called slope aspect.

Assimilation. (1) The incorporation of inorganic or organic substances into cell constituents. (2) The ability of a body of water to purify itself of pollutants.

Assimilative capacity. The capacity of a natural body of water to receive wastewaters or toxic materials without deleterious effects and without damage to aquatic life or humans who consume the water.

Assimilatory nitrate reduction. Conversion of nitrate (NO_3^-) to reduced forms of nitrogen, generally ammonium (NH_4^+), for the synthesis of amino acids and proteins.

Astronomical unit (AU). The average distance from the Earth to the sun. The actual figure for an astronomical unit yielded by this definition, and by the most modern measurements, is approximately 92,955,807 miles (149,597,870.691 km). This figure was adopted in 1996, and is considered accurate to within about 10 feet (approximately 3 m). Also, the distance light travels in a little over eight minutes.

Atmosphere. The sum total of all the gases surrounding the Earth, extending several hundred kilometers above the surface in a mechanical mixture of various gases in fluid-like motion. The permanent constituents are molecular nitrogen; 78.1%, molecular oxygen; 20.9%, argon; 0.934%, and approximately 0.036% carbon dioxide. Various other components exist in trace amounts.

Atmosphere, soil. Gases occupying the pore space in soil. Generally characterized as having a greater percentage of carbon dioxide and a lesser percentage of oxygen than the overlying air.

Atom. The smallest particle of an element, such as nitrogen or hydrogen, composed of electrons and a nucleus (containing protons and neutrons), into which a chemical element can be divided and still retain its characteristic properties. Atoms combine to form ions, e.g., ammonium (NH_4^+), nitrate (NO_3^-), and molecules such as ammonia (NH_3). Atoms are chemical building blocks, and combine with other atoms to form compounds and/or molecules. Atoms rarely exist alone.

Examples: The element hydrogen (H) has the smallest and lightest weight atom of all, an atomic weight close to 1. The element nitrogen (N) has atomic weight of 14, and, therefore, an atom of nitrogen is 14 times heavier than an atom of hydrogen. [also, see *element*]

Atomic number. The number of protons in an element in the nucleus of an atom. [Examples: The atomic number of hydrogen is 1; the atomic number of carbon is 6; the atomic number of calcium is 20].

Atomic weight (relative atomic mass). (1) The sum of the weights of an atom's protons and neutrons, the atomic weight differs between isotopes of the same element. (2) The relative weight of one atom of a given element. [Examples: nitrogen = 14.0067, oxygen = 15.9994, calcium = 40.08]

ATP. See *adenosine triphosphate (ATP)*.

Autotroph. Any organism that is able to manufacture its own food from inorganic substances in its environment. Autotrophs may be photoautotrophic, using light energy to manufacture food, or chemoautotrophic, using chemical energy. [Example: most higher plants are autotrophs, as are algae and many bacteria]. [also see *heterotroph*]

Autotrophic bacteria. See *bacteria, autotrophic bacteria*.

Autotrophic nitrification. The oxidation of ammonium (NH_4^+) to nitrate (NO_3^-) through the combined action of two chemoautotrophic organisms, one forming nitrite (NO_2^-) from ammonium and the other oxidizing nitrite to nitrate. [also see *nitrification*]

Autotrophic organism. (1) Refers to organisms that synthesize their nutrients and obtain their energy from inorganic raw materials. (2) An organism that utilizes carbon dioxide (CO_2) as the sole source of carbon and obtains its energy from the sun (Example: higher plants and algae), or by oxidizing inorganic substances such as sulfur, hydrogen, ammonium, and/or nitrate salts [Example: various bacteria]. [also see *heterotrophic* and *autotrophic bacteria*]

Auxiliary soil and plant substance. Any chemical or biological substance or mixture of substances distributed to be applied to soil, plants, or seeds for soil corrective purposes; or which is intended to improve germination, growth, yield, product quality, reproduction, flavor, or other desirable characteristics of plants; or which is intended to produce any chemical, biochemical, biological, or physical change in soil. Does not include commercial fertilizers, agricultural minerals, soil amendments, or manures. Includes all of the following:

- synthetic polyelectrolytes
- lignin or humus preparations
- wetting agents to promote water penetration
- bacterial inoculants
- microbial products
- soil binding agents
- biotics

[also see *minerals, agricultural; fertilizers, commercial; biotics; amendments;* and *manures*]

Auxin. A group of hormones involved in controlling plant growth and other functions; once thought responsible for phototropism by causing the cells on the shaded side of a plant to elongate, thereby causing the plant to bend toward the light.

Available nutrient (element). That portion of any nutrient (element or compound) in the soil that can be readily absorbed by growing plant roots (in amounts sufficient to plant growth). Same meaning as plant-available. In each case all that is soluble in water is available. In addition, however, some of each that is not soluble in water is available to plants. In general, a form of nutrient capable of being assimilated by a growing plant. [also see *assimilation*]

Available phosphate (P_2O_5). On a fertilizer label with N P & K, the P stands for the sum of the water-soluble and the citrate-soluble phosphate (P_2O_5) in a fertilizer. [also see *water soluble potash (K_2O)*]

Available water (in soil). The portion of water in a soil that can be readily absorbed by plant roots (at rates significant to their growth). The amount of water released between the field capacity and the permanent wilting point. Same as plant-available water.

Available water capacity. The weight percentage of water that a soil can store in a form available to plants. It is equal to the moisture content at field capacity minus that at the wilting point.

"B" horizon, soil. A soil horizon usually beneath the A horizon that is characterized by one or more of the following: (a) a concentration of silicate clay, iron and aluminum oxides, and humus, alone or in combination; (b) a blocky or prismatic structure, and (c) coating of iron and aluminum oxides that give darker, stronger, or redder color. [also see *A horizon*]

B. S. See *base saturation (percentage)*.

Backslope. The hillslope position that forms the steepest, and generally linear, middle portion of the slope. In profile, backslopes are bounded by a convex shoulder above and concave footslope below. [also see *catina; shoulder; footslope;* and *toeslope*]

Bacteria (singular bacterium). A group of unicellular microscopic prokaryotic (where DNA is not organized into chromosomes or surrounded by a nuclear membrane, but is a simple strand in the cell) plants, sometime aggregated into filaments. Found everywhere including soil and water; and parasitic on plants and animals.

Heterotrophic bacteria (using organic compounds as its source of carbon). Bacteria that cannot manufacture organic compounds and so must feed on organic materials that have originated in other plants and animals. Heterotrophic bacteria, together with fungi and a few other groups of heterotrophic organisms, are the decomposers of the biosphere. [also see *fungi*]

Heterotrophic bacteria tend to decompose simpler organic compounds, such as root exudates and fresh plant residue, while fungi tend to decompose more complex compounds such as fibrous plant residues and wood.

> **Photoheterotrophs:** Obtain energy from sunlight and carbon from organic matter
> **Chemoheterotrophs:** Obtain energy and nutritive carbon both from organic matter

Autotrophic bacteria. Bacteria that obtain their energy from sunlight or by the oxidation of inorganic compounds and their carbon by the assimilation of carbon dioxide.

> **Photoautotrophs:** Energy needed for growth and biosynthetic reactions is derived from sunlight and carbon from carbon dioxide. [While not bacteria, green plants are photoautotrophs].
> **Chemoautotrophs:** Obtain energy needed from growth and biosynthetic reactions from the oxidation of inorganic molecules such as nitrogen sulfur, and iron compounds, or from the oxidation of gaseous hydrogen; and carbon from carbon dioxide. [Examples: *Nitrosomonas* (which oxidizes ammonium to nitrite); *Nitrobacter* (which oxides nitrite to nitrate); and *Thiobacillus* (which oxidizes elemental sulfur to sulfates).
>
> The oxidation of elemental sulfur (S) is show below:

$$2S + 2H_2O + 3O_2 \xrightarrow{\textit{Thiobacillus}} 4H^+ + 2SO_4^{2-} \text{ (sulfate)}$$

Banding fertilizer. See *fertilizer, banding*.

Bark. The outer layer of the stems of woody plants; composed of an outer layer of dead cells (cork) and an inner layer of phloem.

Basalt. Highly mafic, igneous, volcanic rock, typically fine-grained and dark in color.

Base. (1) A compound that reacts with an acid to form a salt. (2) A compound that produces hydroxide ions (OH⁻) in aqueous solution. (3) A molecule or ion that captures (takes up) hydrogen ions in a reaction. (4) A molecule or ion that donates an electron pair to form a chemical bond. (5) Also refers to the cations, especially calcium, magnesium, sodium and potassium, which form alkali compounds. A solution of a base has a pH greater than 7. [also see *alkali*]

Base exchange. The replacement of basic cations (Ca, Mg, Na, and K) held on the soil cation exchange complex, by other basic cations. [also see *cation exchange;* and *cation exchange capacity*]

Base fertilizer. See *basic fertilizer*.

Base flow. Stream-flow coming from ground-water seepage into a stream.

Base saturation (percentage) [basic cation saturation]. The extent to which the adsorption complex of a soil is saturated with basic cations (calcium, magnesium, sodium and potassium) [cations other than hydrogen and aluminum]. Base saturation is usually expressed in percentage of the total cation exchange capacity.

Basic cations. Primarily refers to the ions of calcium (Ca^{2+}), magnesium (Mg^{2+}), sodium (Na^+) and potassium (K^+), but also all other cations other than hydrogen (H^+) and aluminum (Al^{3+}).

Optimum Basic Cation Saturation		
	Recommended (%)	Recommended (ppm)
Calcium	80 or greater	Greater than 2000
Magnesium	Less than 10	Less than 400
Potassium	Greater than 5	Greater than 150
Sodium	Less than 5	Less than 150

Compiled From Various Sources

Basic cations are derived from bases including sodium hydroxide (NaOH) [Example: sodium (Na^+)], calcium carbonate ($CaCO_3$) [Example: calcium (Ca^{2+})], and potassium hydroxide (KOH) [Example: potassium (K^+)].

Basic fertilizer. A fertilizer that, after application to and reaction with soil, decreases residual acidity and increases soil pH. [Example: sodium nitrate]. [also see *acid-forming fertilizer;* and *non-acid forming fertilizer*]

Basic rock. An igneous rock that contains less than 55% silica.

Basic slag. A byproduct of steelmaking. Basic slag is largely limestone or dolomite which has absorbed phosphate from the iron ore being smelted. Because of the slowly-released phosphate content, as well as for its liming effect, it is valued as fertilize (mainly in steelmaking areas). According to the American Association of Plant Food Control Officials, basic slag must contain at least 12% total P_2O_5 or be labeled "low phosphate." The available phosphate content of most slag is in the range of 8 to 10%.

Basic solution. An aqueous solution containing more hydroxide (OH^-) ions than hydrogen (H^+) ions. An aqueous solution with a pH greater than 7.

Basicity, residual. The ultimate basicity that develops from fertilizer in a particular soil horizon after the residual salts are removed from that horizon by leaching.

Basidiomycetes. The club fungi, a major group of fungi that all produce a structure (basidium) on which basidiospores are produced. Basidiomycetes include mushrooms and toadstools.

Batch composting. Composting where all materials are processed at the same time without introducing new feedstock once the composting has begun. [Example: windrow systems are batch composting systems].

Bedrock. The solid rock at the surface of the Earth or at some depth beneath the soil and superficial deposits.

Bentonite (hydrated aluminum silicate). Bentonite (general formula (Na, Ca)(Al,Mg)$_6$(Si$_4$O$_{10}$)$_3$(OH)$_6 \cdot$nH$_2$0) is a naturally occurring hydrated aluminum silicate that is useful as a suspending agent in preparation of suspension fertilizers.

Sodium bentonite, a natural clay, has the characteristic of swelling many times its dry size when it becomes wet. When sodium bentonite is applied in a layer of porous soil or mixed with a porous soil and moistened with water, it forms an impermeable layer. Because of this, it is often used as a pond sealer.

Best management practices (BMPs). The management practices determined to permit the least pollution or other undesirable effects. BMPs are those practices that have been proven in research and tested through grower implementation to give optimum production potential, input efficiency, and environmental protection.

Bicarbonate (HCO_3^-). Bicarbonate (chemically referred as the hydrogen carbonate ion) is an anion with the formula HCO_3^- and possess a molecular mass of 61.0. With irrigation water, levels of bicarbonate above 3.0 milliequivalents per liter (meq L^{-1}) are considered harmful since bicarbonates (and associated carbonates [CO_2^{-3}]) will combine with calcium to form lime ($CaCO_3$) when the water evaporates. This results in several negative consequences: (1) when free lime forms, any available beneficial calcium will be precipitated out, further compounding the common problem of not having enough calcium in the soil; and (2) <u>bicarbonate itself is the most toxic anion that exists in relation to plant health</u>.

Bicarbonate toxicities generally arise from deficiencies of iron or other micronutrients caused by the resultant high pH. Two procedures are used to treat irrigation with high levels of bicarbonate. The standard treatment is to lower the water's pH by adding an acid. Lowering the pH to 6.5 neutralizes about half the bicarbonate in irrigation water. Also, the addition of solution-grade calcium sulfate (anhydrite and/or gypsum) to irrigation water will replace any calcium precipitated as lime. The chemical reaction of a bicarbonate ion combining with a calcium ion forming lime (and, thus, tying up soluble calcium):

$$Ca^{2+} + 2H_2CO_3^- \text{ (bicarbonate)} \longrightarrow CaCO_3 \text{ (lime)} + H_2O + CO_2 \text{ (carbon dioxide)}$$

> **Note:** 100-ppm of anything in irrigation water amounts to 270 lbs. per acre-foot of water. Example: A bicarbonate level of 250-ppm (very common in California and other parts of the world) delivers 675 lbs. of bicarbonate per acre foot of water. Many crops may need 2.5 acre-feet of water, so a grower could deliver a surprising 1,688 lbs. of bicarbonate per acre into the soil each year. Each pound of bicarbonate ties up one pound of soluble calcium.

Ag Suitability Water Analysis Bicarbonate Interpretation		
Degrees of Problem [milliequivalents per liter (meq L^{-1})]		
None	Increasing	Severe
less than 1.5	1.5 – 8.0	greater than 8

Biennial. A plant that normally requires two growing seasons to complete its life cycle, flowering and fruiting in its second year. Biennials that are grown for edible leaves or roots are grown as annuals, e.g., beets, Brussels sprouts, cabbage, carrots, celery, lettuce, parsley, and Swiss chard. [also see *annual;* and *perennial*]

Bioaerosols. Airborne particles of biological origin including bacteria, viruses, fungi, yeasts, pollens and organic matter.

Bioassay. Appraisal of the biological activity of a substance by testing its effect on an organism and comparing the result with some agreed standard. Bioassays are usually laboratory tests using a biological test organism.

Bioavailable. (1) Available for biological uptake. (2) Degree of ability to be absorbed and ready to interact in organism metabolism.

Biochemical oxygen demand (BOD). (1) The amount of dissolved oxygen consumed in five days by biological processes breaking down organic matter. (2) The amount of oxygen used in the biochemical oxidation of organic matter. An indication of compost maturity and a tool for studying the composting process.

Biochemistry, soil. The branch of soil science concerned with enzymes and the reactions, activities, and products of soil microorganisms.

Biodegradable. Subject to degradation and/or decomposed by natural enzymatic biochemical processes and activity.

Biodiversity. The variability among living organisms on the Earth, including the variability within and between species and within and between ecosystems.

Biofertilizers. Biological organisms that are involved in fixing N_2 from the air into plant available N, or in increasing the availability of other plant nutrients.

Biogeochemistry. Study of microbially mediated chemical transformations of geochemical interest, such as nitrogen or sulfur cycling.

Biological nitrogen fixation. See *dinitrogen fixation (also called biological nitrogen fixation (BNF)*.

Biological oxidation. The process by which living organisms, in the presence of oxygen, convert organic matter into a more stable or a mineral form.

Biological oxygen demand (BOD). See *biochemical oxygen demand*.

Biomass. (1) The total amount of living organisms and their residues in a volume or mass of the environment. (2) The complete dry weight of organic material found in the biosphere or less strictly, the matter in the biosphere that is contained in living organisms.

Bioremediation. The reduction or elimination of a pollutant by the use of microorganisms.

Biosolids. (1) The resides of wastewater treatment, formerly called sewage sludge. (2) Nutrient-rich organic solid materials produced by sewage sludge and wastewater treatment processes that can be beneficially recycled for plant nutrient content and soil amending characteristics.

Biosphere. (1) The zone of air, land, and water at the surface of the Earth that is occupied by organisms. (2) The environment in which living organisms live. Includes the Earth's crust, vegetation, and atmosphere near the Earth. (3) All ecosystems on Earth as well as the Earth's crust, waters, and atmosphere on and in which organisms exist; also, the sum of all living matter on Earth.

Biosynthesis. The production of needed cellular constituents from other, usually simpler, molecules.

Biotic. Relating to life.

Biotics. All materials for which claims are made relating to organisms, enzymes, or organism by-products.

Biotrophic. Nutritional relationship between two organisms in which one or both must associate with the other to obtain nutrients and grow.

Bioturbation. The disturbance of sediment by organisms, e.g., burrows, trails, or complete mixing.

Biuret (Carbamylurea) [$H_2NC(O)NHC(O)NH_2$]. A compound formed by thermal decomposition or condensation of two molecules of urea. Biuret, a problematic impurity in urea-based fertilizers, is toxic to some crops and should be avoided in the manufacture of fertilizer urea. [also see *urea*]

Black alkali soil. A soil with a high pH (pH = 7.5 to 10.0), exchangeable sodium percentage greater than 15%, and chemistry dominated by sodium carbonate (Na_2CO_3) and/or sodium bicarbonate ($NaHCO_3$). The soils are deflocculated and impervious, and dissolved organic matter/humic substances may be deposited on the soil surface as soil water evaporates. [also see *deflocculation, soil* and *saline-sodic soil*]

Blocky soil structure. Many sided soil aggregates with angular or rounded corners, used for describing soil peds. Aggregates with equidimensional shapes.

Blood meal (dried blood). Blood meal (or dried blood) is the collected blood of slaughtered animals, dried and ground and containing not less than twelve percent (12%) nitrogen (AAPFCO). Nitrogen in blood meal quickly becomes plant available.

Blue-green alga. *See cyanobacteria.*

Bog. Waterlogged, spongy ground; consisting primarily of mosses and containing acidic, decaying vegetation such as sphagnum, sedges, and heaths; which may develop into peat.

Bone meal (6-12-0). Raw bone meal is cooked bones, ground into a meal without any of the gelatin or glue removed. Steamed bone meal has been steamed under pressure to dissolve out part of the gelatin. Bone meal contains calcium phosphate-carbonate [$3Ca_3(PO_4)_2 \cdot CaCO_3$]. The guaranteed analysis for cooked bone meal is typically: total nitrogen (N) = 6%; available water-soluble nitrogen = 1%; water-insoluble nitrogen = 5%; available P2O5 = 12%; calcium = 15%.

Borax ($Na_2B_4O_7 \cdot 10H_2O$). A white crystalline salt used in fertilizer as a source of the plant nutrient element boron. Fertilizer borax contains approximately 11 percent boron, while pure borax contains 11.34% boron.

Boron (B). An essential non-metallic micronutrient that exists in the soil in a number of primary and secondary nutrients. Organic matter is also an import source of soil boron. Boron is absorbed by plants as boric acid (H_3BO_3) or one of the borate anions. Boron is essential for germination of pollen grains and growth of pollen tubes, and is essential for seed and cell wall formation and affects protein formation. Boron deficiency generally results in stunted plant growth with the growing point and younger leaves first due to lack of mobility in the plant.

In crops, the symptoms of boron deficiency exists as crooked and cracked stem in celery, corky core in apples, black heart in beets and ringed or banded leaf petioles in cotton. In alfalfa, rosetting (yellow top) followed by death of the terminal bud due to low boron concentration in the plant. Coarse-textured sandy soils low in organic matter are typically low in minerals that contain boron and boron availability. High soil pH values also limit boron availability. In California soils are typically low in boron on the east side of the San Joaquin Valley, while toxically higher amounts exists on the west side of the San Joaquin Valley. There is a narrow range between deficiency and toxicity with boron levels in soils. [also see *plant nutrients, essential;* and *anion*]

Bottomland (obsolete). An obsolete, informal term loosely applied to varying portions of a flood plain.

Brackish. Slightly salty. Water with a content of 1.5% to 3% (1,500 to 5,000 ppm) salts. Seawater has more than 3% salts. [also see *sea/ocean water analysis;* and *saline water*]

Breccia. A coarse-grained, clastic rock (similar to a conglomerate) composed of angular rock fragments (larger than 2 millimeters) commonly bonded by mineral cement in a finer-grained matrix of varying composition and origin. Breccia is the consolidated equivalent of rubble. [also see *conglomerate*]

Brimstone. Sulfur (S) [as opposed to the torments of damnation to hell; as in hellfire and brimstone (see Rev. 20:10)]

Broadcast. Seed, fertilizer and soil amendments uniformly spread on the soil surface. They may incorporated into the soil.

Brown algae. See *algae, brown*.

Buffer, buffering capacity of soils. (1) The capacity of the soil solids and liquids to resist appreciable change in pH (changes when limited amounts of acid or base are added) of the soil solution. (2) The ability to maintain the approximate concentration desired of any ion in the soil solution. In soils, organic matter, clays, and free calcium carbonate tend to buffer the system again pH changes. [Examples: clays, humus, and free calcium carbonates. Carbonate ions (CO_3^{2-}) are a buffer as they react with hydrogen ions (H^+) to form bicarbonate ions (HCO_3^-)].

Buffer solutions. Aqueous solutions consisting of a mixture of a weak acid and its conjugate (noting two or more liquids in equilibrium with one another) base, or a weak base and its conjugate acid. These solutions have the property that the pH of the solution changes very little when a small amount of strong acid or base is added to it. Buffer solutions are used as a means of keeping pH at a nearly constant value in a wide variety of chemical applications. Many life forms thrive only in a relatively small pH range. An example of a buffer solution is blood.

Buffers. Chemicals that maintain pH values within narrow limits by absorbing or releasing hydrogen ions. Buffers are usually a mixture of weak acids and their salts which tend to resist changes in pH. In soils, organic matter (esp. humus), clays and free calcium carbonate tend to buffer the system against pH changes.

Bulk density (soil). The mass (or weight) of dry soil per unit bulk volume, including the air space. In general expressed in grams per cubic centimeter. [Note: Research has shown that corn roots are unable to penetrate a layer of soil having a bulk density of

1.85 g cm^{-3}]. Bulk density of soils is altered considerably by such factors as moisture content, degrees of compaction, degrees of aeration and parent material. [Example: if the top 6 inches of a given acre of soil weighs exactly 2 X 10^6 (2 million) pounds, the bulk density of that soil is 1.47 grams per cubic centimeter (g cm^{-3}), or 91.83 pounds per cubic foot]. [also see *particle density, soil*]

Expected Range for Bulk Density Values (g cm^{-3})	
Most soils	1.0 – 1.8
Peat (organic soils)	0.05 - 0.5
Sandy soils	1.4 – 1.7
Clayey soils	1.0 – 1.5
Compacted soils	1.6 – 2.0

Compiled From Various Sources

Bulking agent. Material, usually carbonaceous such as sawdust or woodchips, added to a compost system to maintain airflow by preventing settlement and compaction of the waste.

"C" horizon, soil. A mineral horizon generally beneath the A and B horizons (solum) that is relatively unaffected by biological activity and pedogenesis; and is lacking properties diagnostic of an A or B horizon.

Calcareous (soil). Soil with sufficient carbonates (mostly calcium but also magnesium) to effervesce (or fizz) visibly with cold 0.1 *N* hydrochloric acid (HCl). In soil science, especially refers to the presence of calcium carbonate, lime, (CaCO$_3$) which acts as a buffer to prevent pH changes below 7.6.

Calcification. The processes of calcium carbonate accumulation.

Calcination (calcining). A thermal high-temperature treatment process applied to ores and other solid materials/substances in order to bring about a thermal decomposition, phase transition, or removal of a volatile moisture fraction. The calcination process normally takes place at temperatures below the melting or fusing point of the products or materials. Calcination is to be distinguished from roasting, in which more complex gas-solid reactions take place between the furnace atmosphere and the solids. [also see *hemihydrate*]

Calcite. The crystalline natural mineral calcium carbonate ($CaCO_3$). Calcite crystallizes in the hexagonal system, with the main types of crystals in soils being "dog-tooth," prismatic, nodular, fibrous granular, and compact. Calcite is the basic constituent of limestone, marble, and chalk.

Calcite can be either dissolved by groundwater or precipitated by groundwater, depending on several factors including the water temperature, pH, and dissolved ion concentrations. Although calcite is fairly insoluble in cold water, acidity can cause dissolution of calcite and release of carbon dioxide gas. Calcite, like most carbonates, will dissolve with most forms of acid.

Calcitic lime. The rocks limestone or marble containing mostly the mineral calcite ($CaCO_3$). [also see *lime, agricultural*]

Calcium (Ca). Calcium, comprising about 3.6% of the Earth's surface, is unique among the required plant elements in that it is both an essential plant nutrient as well as a soil amendment. As a plant nutrient, calcium is absorbed by plants as Ca^{2+} from the soil solution, and is supplied to the root surface by mass flow and root interception. The normal concentration in plant leaves ranges between 0.2% to 1.0%. The ability for plants to uptake Ca^{2+} is limited because it can be absorbed only by young root tips in which the cell walls of the endodermis are still unsuberized. Conditions impairing the growth of new roots will reduce access of plant roots to Ca^{2+} and induce deficiency. Problems related to inadequate Ca^{2+} uptake are more likely to occur with plants that have small root systems than with those possessing more highly developed rooting systems.

Because calcium is required for cell division and elongation as well as cell wall strength, deficiencies first appear at root tips and other growing points. Weakened stems, premature shedding of blossoms and buds are all symptoms of calcium deficiency. Blossom-end rot in tomatoes and peppers; bitter pit and cork spot in apples; club root in broccoli, cauliflower and other cole crops; cavity spot in carrots; tip burn of lettuce and cabbage; "pillowy" fruit disorder in cucumber; blackheart in celery; internal brown spot in potatoes are also all the result of calcium deficiencies within the plants. Low calcium also reduces fruit quality and seed formation and quality in all crops.

As a soil amendment calcium helps (a) improve soil structure, (b) improve, amend and reclaim soils high in destructive sodium, (c) replace harmful salts such as sodium, (d) improve water use efficiency, (e) reduce runoff and erosion, (f) prevent soil crusting, (g) improve compacted soils, and (h) prevent water-logging of soils. The addition of calcium to irrigation water will also help replace any calcium precipitated as lime due to high concentrations of bicarbonate. [also see *plant nutrients, essential*; *amendments, soil*; *infiltration*; and *cation*]

Calcium ammonium nitrate (CAN) [$5Ca(NO_3)_2 \cdot NH_4NO_3 \cdot 10H_2O$]. A white or grey granular which dissolves in water absolutely. CAN is a new type high-efficient compound fertilizer that contains nitrogen and calcium and also supplies nitrogen to plant quickly since it and can be absorbed by the plant directly. CAN is a neutral fertilizer. Total nitrogen: 15.5% minimum, calcium 18.5%.

Calcium carbonate. See *calcite, limestone* and *lime, agricultural*.

Calcium carbonate equivalent. (1) The content of carbonate in a liming material or calcareous soil calculated as if all of the carbonate is in the form of the salt $CaCO_3$. (2) The amount of calcium carbonate required to neutralize the acidity produced by a given quantity of fertilizer product. [also see *lime, agricultural*]

Calcium cyanamid ($CaCN_2$). An organic material containing approximately 21 percent nitrogen. It is alkaline in reaction and is used not only as a fertilizer, but also as a defoliant and for the control of weeds and certain soil-borne diseases.

Calcium nitrate ($Ca(NO_3)_2$). A fertilizer that is chiefly the calcium salt of nitric acid. It should contain not less than fifteen (15) percent nitrate nitrogen.

Calcium phosphate. See *monocalcium phosphate. ($Ca(H_2PO_4)_2$.*

Calcium sulfate ($CaSO_4$). See *anhydrite, gypsum,* and *hemihydrate*.

Caliche. A general term for a prominent zone of secondary carbonate accumulation in surface materials in warm, subhumid to arid areas. Caliche is formed by both geologic and pedologic processes. Finely crystalline calcium carbonate forms a nearly continuous surface-coating and void-filling medium in geologic (parent) materials. Cementation ranges from weak in non-indurated forms to very strong in types that are indurated. Other minerals (carbonates, silicate, sulfate) may be present as accessory cements. Most petrocalcic and some calcic horizons are caliche.

Calvin cycle. (aka Calvin-Benson Cycle or Carbon Fixation). A series of biochemical, enzyme mediated reactions during which atmospheric carbon dioxide is reduced and incorporated into organic molecules. Eventually some of the reduced carbon dioxide forms sugars. In eukaryotes, this occurs in the stroma (the spongy, colorless matrix of a cell that functionally supports the cell) of the chloroplast.

CAN. See *calcium ammonium nitrate (CAN)*.

Canopy. Layer of vegetation elevated above the ground, usually of tree branches and epiphytes.

Capillarity. The process by which moisture moves in any direction through the fine pore spaces and as films around particles.

Capillary action. The means by which liquid moves through the porous spaces in a solid, such as soil and plant roots, due to the forces of adhesion, cohesion, and surface tension. Capillary action is essential in carrying substances and nutrients from one place to another in plants and animals.

Capillary fringe. The zone just above the water table that remains practically saturated with water.

Capillary moisture. That amount of water that is capable of movement after the soil has drained. Capillary moisture is held by adhesion and surface tension as films around soil particles and in the finer pore spaces.

Carbamylurea. See *Biuret (Carbamylurea) [$H_2NC(O)NHC(O)NH_2$]*.

Carbohydrate. An organic biochemical compound, $(CH_2O)_n$, consisting of a chain of carbon atoms to which hydrogen and oxygen are attached in a 2:1 ratio. Carbohydrates are generally easily assimilated by soil microorganisms, and serve as energy sources and structural materials for cells of all organisms. [Examples: sugars, starch, chitin, steroids, and cellulose]

Carbon (C). Carbon is an essential plant nutrient that forms the skeleton for all organic molecules. Therefore, it is a basic building block for all plant life. Carbon is assimilated from the atmosphere by plants in the form of carbon dioxide (CO_2) through the process of photosynthesis. In this process, carbon is combined with hydrogen and oxygen to form carbohydrates. Further chemical combinations, some with other essential elements, produce the numerous substances required for plant growth. [also see *plant nutrients, essential*]

Carbon-14. An isotope of carbon-12 (containing two more neutrons) that is radioactive and used in carbon dating.

Carbon cycle. The sequence of transformations where carbon dioxide is converted to organic forms, generally by photosynthesis, recycled through the biosphere (with partial incorporation into sediments), and ultimately returned to its original state through respiration or combustion.

Carbon dioxide (CO_2). A compound consisting of one carbon and two oxygens. Carbon dioxide is a reactant in photosynthesis and is necessary for plant life. CO_2, a trace gas, comprises 0.039% of the atmosphere, whereas in aerobic soils it is higher in the order of 0.2 to 1%. By comparison, oxygen (O_2), an essential plant nutrient, is found in aerobic soils in about 20.3% by volume as compared with that of 20.95% in the atmosphere. CO_2 is released in the soil by means of ammonification of organic matter as shown in the chemical formula below:

$$\text{organic compounds} + O_2 \xrightarrow{\text{enzymes (from bacteria)}} CO_2 + H_2O + \text{energy} + \text{nutrients (e.g., } NH_4^+\text{)} + \text{humus}$$

[also see *ammonification*]

Carbon fixation. Conversion of carbon dioxide or other single-carbon compounds to organic forms such as carbohydrates.

Carbon:nitrogen ratio. The ratio of the weight (mass) of organic carbon to the weight of total nitrogen in the soil or in organic material. Generally when the C:N ratio is greater than 25:1, net immobilization occurs, while with ratios less than 25:1 net mineralization is likely. [also see *immobilization; mineralization;* and *priming effect*]

Carbonate. (1) A mineral composed mainly of calcium (Ca^{2+}) and carbonate (CO_3^{2-}) ions [$CaCO_3$], but may also include magnesium (Mg^{2+}) [$MgCO_3$], iron (Fe^{3+}) [$Fe_2(CO_3)_3$] and others. (2) Rock or sediments derived from debris of organic materials composed mainly of calcium and carbonate (e.g., shells, corals, etc.) or from the inorganic precipitation of calcium (and other ions) and carbonate from solution (seawater). (3) The bicarbonate (HCO_3^-) and carbonate (CO_3^{2-}) content of irrigation water or soil may also cause precipitation of calcium and magnesium carbonates, e.g., limestone or dolomite, and increase the SAR of the soil. [also see *bicarbonate (HCO_3^-)*]

Carbonic acid (H_2CO_3). The inorganic weak acid compound with the formula H_2CO_3. Carbonic acid is also a name sometimes given to solutions of carbon dioxide dissolved in rainwater which contain small amounts of H_2CO_3. The salts of carbonic acids are called bicarbonates (or hydrogen carbonates) and carbonates.

Chemical weathering is the breakdown of rocks into sediment by chemical processes. One of the main agents of chemical weathering of rock and minerals is carbonic acid. Rainwater and groundwater are not pure water; some of the molecules of water react with the carbon dioxide in the atmosphere (in the case of rainwater) or produced by bacteria and plant roots (in the case of groundwater) producing carbonic acid, as follows:

$$H_2O + CO_2 \longrightarrow H_2CO_3 \text{ (carbonic acid)}$$

Casing layer. In mushroom farming, a moist layer of organic material (often peat moss) with a small amount of calcium carbonate (agricultural lime) or calcium sulfate (anhydrite or gypsum) that growers apply over mycelium to retain moisture and provide a growing surface. [also see *anhydrite*; *gypsum*; and *lime, agricultural*]

Casparian strip. In plant anatomy, the casparian strip is a band of cell wall material in the radial and transverse walls of the endodermis, which is chemically different from the rest of the cell wall. The casparian strip blocks the passive flow of materials, such as water and solutes, into the stele of a plant. [also see *stele*]

Cat-clay (obsolete). See *acid sulfate soils*.

Catabolic reactions. Reactions in cells in which existing chemical bonds are broken and molecules are broken down. These reactions generally produce energy, involve oxidation, and lead to a decrease in atomic order. [also see *catabolism*]

Catabolism. (1) Biochemical processes involved in the breakdown of organic compounds, usually leading to the production of energy. (2) Collectively, the chemical reactions resulting in the breakdown of complex materials and involving the release of energy. The energy-releasing breakdown of molecules. [also see *catabolic reactions*; *metabolism*; and *anabolism*]

Catalyst. (1) A substance that increases the rate of reaction without being consumed in the overall reaction. (2) A substance that promotes a chemical reaction by lowering the activation energy without itself being changed in the reaction. [Example: enzymes are a type of catalysts]

Catchment. (1) A catching or collecting of water, especially rainwater. (2) A structure, such as a basin or reservoir, used for collecting or draining water. (3) The amount of water collected in such structures. (4) A catchment area.

Catena. A sequence of soils across a landscape of about the same age, derived from similar parent material, and occurring under similar climatic conditions; but having different characteristics due to variations in relief and in drainage. [Example: the summit of a hill with its *shoulder, backslope, footslope, and toeslope*]

Cation. Ions with one or more atoms with a positive electrical charge such as ammonium (NH_4^+), calcium (Ca^{2+}), or zinc (Zn^{2+}). A charged form of an atom or molecule carrying one or more positive charges of electricity (valence). Metals typically form cations. [also see *anion*]

Plant Nutrients and Other Important Cations (in Soils and Irrigation Water)		
Element Name	Chemical Symbol	Ionic Forms
Aluminum	Al	Al^{3+}
Cadmium	Cd	Cd^{2+}
Calcium	Ca	Ca^{2+}
Cobalt	Co	Co^{2+}
Copper	Cu	Cu^{2+}

Hydrogen	H	H+
Iron	Fe	Fe^{3+} (ferric) [oxidized] Fe^{2+} (ferrous) [reduced]
Lead	Pb	Pb^{2+}
Magnesium	Mg	Mg^{2+}
Manganese	Mn	Mn^{2+}
Mercury	Hg	Hg^{2+}
Nickel	Ni	Ni^{2+}
Nitrogen	N	NH_4^+
Potassium	K	K^+
Silicon	Si	Si^{4+}
Sodium	Na	Na^+
Zinc	Zn	Zn^{2+}

Cation exchange. The exchange of cations in solution for an adsorbed (held) cation at the negatively charged surfaces of soil clay and organic matter particles. Cation exchange is an important reaction in soil fertility in correcting soil acidity and alkalinity, in changes altering soil physical properties, and as a mechanism in purifying or altering percolating waters. [also see *exchangeable ion;* and *base exchange*]

Cation exchange capacity (CEC): The sum total of exchangeable cations that a soil can adsorb, often expressed in terms of milliequivalents per 100 grams (meq/100g) of dry soil, or centimoles per kilograms dry soil. Sometimes called "total-exchange capacity," "base-exchange capacity," or "cation-adsorption capacity." The sum of exchangeable calcium, magnesium, potassium, sodium, and aluminum generally equals, for all practical purposes, the soil's cation exchange capacity. The higher the CEC, the greater the soil's ability to adsorb cations. [also see *exchangeable ion*].

The CEC measures how many negative soil sites are available for the positive charged cations to adhere to. This is important because the plant nutrients will stick to the soil and not be leached out of the root zone based on the CEC measurements. Clay has total CEC readings of 20 to 30 meq/100g soil. With fewer negative soil sites, sand has a lower total CEC reading of 2 to10 meq/100g of soil.

Cation exchange sites. Locations on the surface of soil colloids (clay, organic matter) with negative charges capable of attracting and holding positively charged cations.

CEC. See *cation exchange capacity*.

Cell. The structural unit of all organisms. In plants, cells consist of the cell wall and the protoplast. The fundamental unit of all living matter.

Cellobiose ($C_{12}H_{22}O_{11}$). A disaccharide obtained by the hydrolysis of cellulose by the enzyme cellulase.

Cellulase. An enzyme that converts cellulose to the disaccharide cellobiose.

Cellulose ($C_6H_{10}O_5)_n$). (1) An organic polysaccharide compound consisting of a linear chain of several hundred to over ten thousand linked glucose units. Cellulose is the main component of plant cell walls and is <u>the most abundant compound on Earth that is manufactured by living things</u>. (2) The chief component of all plants; found mainly in the cell wall in plants and green algae. Cellulose is not easily digested by microorganisms, and is indigestible in the human intestine.

Cell wall. The rigid structure deposited outside the cell membrane. Plants are known for their cell walls of cellulose, as are the green algae and certain protists, while fungi have cell walls of chitin.

Cemented (soil). Soils having a hard, brittle consistency because the separates are held together by cementing substances such as humus, $CaCO_3$, or the oxides of silicon, iron, and aluminum. The hardness and brittleness persist even when wet. [also see *pan*; *duripan*; *hardpan*; and *claypan*].

Certified organic. As defined by many state laws, food can be labeled "organic" or "certified organic" if it has been produced without synthetically compounded fertilizers, pesticides, or growth regulators. For complete information on certified organics and the USDA's National Organic Program (NOP), visit: www.ams.usda.gov/nop and the Organic Materials Review Institute, www.omri.org/. [also see *fertilizer, natural organic*]

Chalk. (1) Soft white limestone which consists of very pure calcium carbonate ($CaCO_3$) and leaves little residue when treated with hydrochloric acid (HCl). Chalk generally consists of the remains of foraminifera, echinoderms, mollusks, and other marine organisms and fossil seashells; with vary amounts of silica, quartz, feldspar, or other mineral impurities. (2) The upper or final member of the cretaceous system.

Channel. A tubular-shaped soil pore.

Chelate. Organic chemicals with two or more functional groups that can bind with metals to form a ring structure. Soil organic matter can form chelate structures with some metals. Inorganic sources of copper, iron, manganese and zinc generally are converted to insoluble compounds soon after soil application. For that reason, artificial chelating compounds are sometimes added to soil to increase the soluble fraction of these metal micronutrients. Among the best chelating agents known are EDTA (ethylenediaminetetraacetic acid), HEDTA (hydroxyethylenediaminetriacetic acid), and DTPA (diethylenetriamine-pentaacetic acid. Eventually chelates are microbiologically degraded. [also see *ligand*]

Chelated plant nutrients. Compounds of metallic secondary and micronutrients which have reacted with organic chelating agents and have the property of being available under soil pH conditions in which the nutrients normally form insoluble compounds (AAPFCO).

Chelating agent (sequestering agent). A compound having two or more sites of attachment to a metal (cation or anion) to form a chelate (AAPFCO).

Chemical formula. A notation that uses atomic symbols with numerical subscripts to convey the relative proportions of atoms of the different elements in the substance. [Examples of chemical formulas: $NaCl$ = table salt; NH_3 = anhydrous ammonia; H_2SO_4 = sulfuric acid; and $CaSO_4 \cdot 2H_2O$ = gypsum]

Chemical reaction. The making or breaking of chemical bonds between atoms or molecules.

Chemical weathering (decomposition). (1) One of the two basic processes of weathering of rocks and minerals into soil. (2) The breakdown of rocks and minerals due to the presence of water and other components in the soil solution, or changes in oxidation and/or reduction. Hydrolysis, hydration, acidification, oxidation, and dissolution are all forms of chemical weathering. [also see *physical weathering*]

Chemigation. Application of pesticides, fertilizers, and amendments by mixing them into irrigation water, usually in sprinkler or drip systems; to fertilize crops and other plants, control pests, or amend soils. [also see *fertigation*]

Chemoautotroph. Organisms that obtain energy from the oxidation of chemical (generally inorganic) compounds and carbon from carbon dioxide.

Chemoheterotroph. Organisms that obtain energy and carbon from the oxidation of organic compounds.

Chemolithotroph. Organisms that obtain energy from the oxidation of inorganic compounds and uses inorganic compounds as electron donors.

Chemotrophs. Organisms (usually bacteria) that derive energy from inorganic reactions; also known as chemosynthetic.

Chiseling. Tillage with an implement having one or more soil-penetrating points that shatter or loosen hard, compacted layers, often to a depth below normal plow depth.

Chitin ($C_8H_{13}O_5N_n$). A carbohydrate polymer found in the cell walls of fungi and in the exoskeletons of arthropods; which provides strength for support and protection.

Chlorine (Cl). Plants utilize this non-metallic essential micronutrient in the form of chloride (Cl^-), the only form in which this element exists in the soil. Chloride is involved in energy reactions in the plant, specifically involved in the chemical breakdown of water in the photosynthesis reaction. It also activates several enzyme systems and is involved in transporting several cations (e.g., potassium, calcium, magnesium) within the plant, regulating the actions of stomatal guard cells, thus controlling water loss and moisture stress while maintaining plant turgor. Chloride also helps diminish the effect of fungal root and leaf diseases in many grass species. Chloride is very mobile in the soil and leaches readily. Deficiencies, while rare, are most likely on sandy soils but can occur with any soil texture. [also see *plant nutrients, essential;* and *anion*]

Chlorophyll. The green-colored pigment that absorbs light during photosynthesis; generally found in plants, algae, and some bacteria. Chlorophyll includes a porphyrin ring and often has a long hydrophobic tail. Magnesium (Mg) is the central ion in the chlorophyll molecule.

chlorophyll a ($C_{55}H_{72}O_5N_4Mg$). The green photosynthetic pigment common to all photosynthetic organisms.

chlorophyll b ($C_{55}H_{70}O_6N_4Mg$). An accessory chlorophyll found in green algae and plants.

chlorophyll c ($C_{35}H_{30}O_5N_4Mg$). An accessory chlorophyll found in some protists.

Chloroplast. (1) Disk-like organelles with a double membrane found in eukaryotic plant cells; containing thylakoids and are the site of photosynthesis. (2) A chlorophyll containing plastid found in algal and green plant cells.

Chlorosis. A loss of normal green color, i.e., chlorophyll, of the plant (particularly the leaves). Chlorosis is a sign of nutrient deficiency, and specific patterns of chlorosis are characteristic of individual nutrients. Chlorotic leaves range from light green through yellow to almost white. [also see *interveinal chlorosis;* and *necrosis*]

Chroma. The relative purity of a color directly related to the dominance of the determining wavelength. One of the three variables of color. Chroma is used to determine the color of soils. [also see *hue;* and *value*]

Chronosequence. A sequence of soils that changes gradually from one to the other with time.

Clay. (1) A mineral (inorganic) soil separate consisting of particles less than 0.002 millimeter in equivalent diameter. (2) A soil textural class. (3) A specific crystalline or amorphous mineral structure. [also see *silt; sand; loam;* and *texture, soil*]

Clayey. Texture groups consisting of sandy clay, silty clay, and clay soil textures. Family particle-size class for soils with 35% or more clay.

Claypan. A dense, compact, slowly permeable layer in the subsoil, with a much higher clay content than overlying materials from which it is separated by a sharply defined boundary. Claypans are usually hard when dry, and plastic and sticky when wet. [also see *pan*; *cemented (soil)*; *hardpan*; and *duripan*]

Cleavage. The ability of a mineral or rock to split along predetermined planes of weakness.

Climax community. The stage in community succession where the community has become relatively stable through successful adjustments to its environment.

Climax vegetation. A fully developed plant community that is in equilibrium with its environment.

Clod. A compact, coherent mass of soil ranging in size from 5 to 250 millimeters (0.2 to 10 inches); produced artificially, usually by such human activities as plowing and digging, especially when these operations are performed on soils that are either too wet or too dry for normal tillage operations. [also see *structure, soil*; and *ped*]

Coated (slow release) fertilizer. A fertilizer containing sources of water-soluble nutrients, release of which in the soil is controlled by a coating applied to the fertilizer.

Coating. A layer of a substance completely or partly covering a surface. Coatings are composed of a variety of substances, separately or in combination, including: clay coatings (skins), calcite coatings, etc. Coatings may become incorporated into the matrix or may be fragmented.

Cobalt (Co). While not universally accepted as an essential plant nutrient, in the form of Co^{2+} cobalt is essential for ribonucleotide reductase in *Rhizobia* bacteria fixing N_2. Cobalt, therefore, is needed in the nodules of legumes, such as beans and clovers. Only 10 parts per billion (ppb) of Co^{2+} in nutrient solution was found to be adequate for N_2 fixation by alfalfa. Cobalt sulfate ($CoSO_4 \cdot 7H_2O$) is sometimes added in mixed fertilizer blends for use on pastures, etc., where the soil is deficient in this element. [also see *plant nutrient essential;* and *cations*]

Co-composting. (1) The process of blending biosolids with manures or other green waste materials to produce compost. (2) Composting processes utilizing carbon-rich organic materials (e.g., leaves, yard wastes, etc.) in combination with nitrogen-rich amendments (e.g., sewage sludges). Co-composting includes both the active and curing phases of the composting process. [also see *compost*]

Cohesion. (1) The mutual attraction of molecules of the same substance. (2) The force that holds molecules of the same substance together. (Example: water molecules held together by other water molecules). Cohesion decreases with rise in temperature. [also see *adhesion*]

Cohesion-adhesion theory. Cohesion is the ability of water molecules to stick together (held by hydrogen bonds), forming a column of water extending from the roots to the leaves. This describes the properties of water that help move it through a plant. Adhesion is the ability of water molecules to stick to the cellulose in plant cell walls, counteracting the force of gravity and helping to lift the column of water.

Collembola. The accurate name for the springtails. An order of wingless insects, having a forked springing organ on the 4th abdominal segment. Collembola live in the surface of the soil and feed on organic matter.

Collenchyma cells. Collenchyma tissues are mainly found under the epidermis in young stems in the large veins of leaves. The cells are composed of living, elongated cells running parallel to the length of organs that it is found in. Collenchyma cells have thick cellulose cell walls which thickened at the corners. Intercellular air spaces are absent or very small. The cells contain living protoplasm and they sometimes contain chloroplasts. [also see *parenchyma cells*]

Colloid. A soil particle (organic or inorganic) with a diameter of 0.1 to 0.001 micrometer (μm) that, when suspended in water, diffuses not at all or very slowly through a semi-permeable membrane. Colloids have a vast surface area per unit mass, which accounts for their high adsorptive capacity and their high cation exchange capacity in soils. [Example: calcium in milk, and clays suspended in muddy water]. [also see *cation exchange*]

Colluvial. Pertaining to material or processes associated with transportation and/or deposition by mass movement (direct gravitational action) and local, unconcentrated runoff (overland flow) on sideslopes and/or at the base of slopes.

Colluvium. Rock fragment deposits and soil material accumulated at the base of steep slopes as a result of gravitational action. [also see *talus*; *scree*; and *angle of repose*]

Colonization. The establishment of a community of microorganisms at a specific site or ecosystem.

Commercial fertilizer. See *fertilizer, commercial*.

Community. All organisms that occupy a common habitat and interact with one another.

Compaction, soil. The changing of loose sediment into hard, firm material. The process by which the soil grains are rearranged to decrease void space and bring them into closer contact with one another, thereby increasing the bulk density and reducing soil porosity, water and air movement, and root growth. Compaction is usually attained by the application of mechanical forces (e.g., tractor wheels) to the soil.

Competition. (1) Rivalry between two or more species for a limiting factor in the environment that usually results in reduced growth of participating organisms. (2) One of the biological interactions that can limit population growth; occurs when two species vie with each other for the same resource.

Complete fertilizer. See *fertilizer, complete*.

Composite structure. Any combination of various types of soil peds.

Compost. (1) A biologically stable material derived from the composting process. Composting is the biological decomposition of organic matter that inhibits pathogens, viable weed seeds, and odors. Composting may be accomplished by mixing, and piling in a way as to promote aerobic or anaerobic decay, or both. (2) Organic residues that have been mixed, piled, and moistened, with or without addition of fertilizer and soil amendments, and generally allowed to undergo thermophilic decomposition until the original organic materials are substantially altered or decomposed. (3) Organic residues or a mixture of organic residues and soil that have been piled and allowed to undergo biological decomposition until the original organic materials have been substantially altered or decomposed. Often called "synthetic manure." Composts are used for fertilizing and conditioning the soil. Composting materials include landscape trimmings, agricultural crop residues, paper pulp, food scraps, wood chips, manures and biosolids. [also see *co-composting*]

Compost extract. A solution produced by draining or leaching water through compost. The solution contains soluble nutrients but very few soil microorganisms. Usually, water is cycled through the compost a number of times in order to extract soil microorganisms from the compost surface. Relatively few microorganisms are extracted with this process as compared to compost tea. [also see *compost tea; manure tea;* and *compost leachate*].

Compost leachate. A solution produced when water drains from over-saturated compost. The leachate typically contains only soluble materials, nutrients, and a few soil microorganisms. Compost leachate often becomes anaerobic during passage through the compost due to the fact that if enough soluble carbohydrates are present in the compost or in the leachate, then bacteria and fungi can grow using available oxygen. Phytotoxic compounds may then be present in the leachate. [also see *leachate; phytotoxic; compost tea; manure tea;* and *compost extract*].

Compost tea. A water extract of compost that is brewed, where the organisms extracted from the compost (bacteria, fungi, protozoa and nematodes) are given a chance to increase in number and activity using the soluble food resources and nutrients present in the water. An enormous diversity of bacteria, fungi, protozoa, and nematodes can be present, depending on the quality of the compost. [also see *manure tea; compost extract;* and *compost leachate*]

Composting. A controlled biological process that converts organic constituents (usually wastes) into stable organic residues suitable for use as a soil amendment or organic fertilizer.

Compound. A material that is a specific combination of atoms or ions, the smallest particle of which is one molecule, held together by covalent or ionic bonds chemical bonds. [Example: NH_3 is the chemical formula for the molecule of the compound ammonia].

Compound soil structure. Large soil peds, such as blocks or prisms, which are themselves composed of smaller, incomplete soil peds.

Concretion. A small, hard, local concentration of material such as calcite, anhydrite, iron oxide, or aluminum oxide. Concretions are usually spherical or sub-spherical but may be irregular in shape.

Condensation. (1) The process of water vapor in the air turning into liquid water. (2) The change of a gas to either the liquid or the solid state. Water drops on the outside of a cold glass of water are condensed water. Condensation is the opposite process of evaporation. [also see *evaporation*]

Conditioner, soil. A conditioner is any material that measurably improves specific soil physical characteristics or physical processes for a given use of as a plant growth medium. Examples of soil conditioners include gypsum, crop residues, animal manures, composts, sewage sludge, and certain cellulose and lignin derivatives. Soil conditioners tend to agglomerate soil colloids and produce a much improved, granular soil structure, thereby increasing the permeability of the soil to air and water and reducing crusting of dry soil. [also see *amendment, soil;* and *gypsum*]

Conductivity. See *electrical conductivity*.

Conglomerate. A coarse-grained, clastic sedimentary rock composed of rounded to subangular rock fragments (clasts) larger than 2 millimeters, commonly with a matrix of sand and finer material. Cements include silica, calcium carbonate, and iron oxides. The consolidated equivalent of gravel.

Conservation, soil. (a) Protection of the soil against physical loss by erosion or against chemical deterioration; that is, excessive loss of fertility by either natural or artificial means. (b) A combination of all management and land use methods that safeguard the soil against depletion or deterioration by natural or by human-induced factors.

Conservation tillage. The U.S. Natural Resource Conservation Service defines conservation tillage as any tillage system that leaves at least 30% of the surface covered by plant residues for control of erosion by water. For controlling erosion by wind, it means leaving at least 1000 pounds per acre small-grain-equivalent during the critical wind erosion period. [also see *conventional tillage*]

Consistence, soil. The resistance of the soil to deformation or rupture as determined by the degree of cohesion or adhesion of the soil particles to each other.

Consistency, soil. The combination of properties of soil material that determine its resistance to crushing and it ability to be molded or changed in shape.

Consortium. Two or more members of a natural assemblage in which each organism benefits from the other. The group may collectively carryout some process that no single member can accomplish on its own.

Consumer. Any organism which must consume other organisms (living or dead) to satisfy its energy needs.

Contamination. Any substance added or accumulating in air, water, or soil that makes the air, water, or soil less desirable for human and animal use, and plant growth.

Conventional tillage. A combination of tillage operations traditionally used for a given region and crop. Any tillage that leaves less than 15% of soil covered with crop residues at the time of planting. Typical conventional tillage may include moldboard plowing, followed by two disking operation, then harrowing and seeding. [also see *conservation tillage*]

Conveyance loss. Water that is lost in transit from a pipe, canal, or ditch by leakage or evaporation. Generally, the water is not available for further use; however, leakage from an irrigation ditch, for example, may percolate to a ground-water source and be available for further use.

Copper (Cu). Copper is an essential metallic micronutrient that is absorbed by plants from the soil in the form of the Cu^{2+} ion. Copper is necessary for chlorophyll formation in plants and catalyzes several other plant reactions, although it is not usually a part of the products formed by those reactions. Organic soils are most likely to be copper deficient since copper is fixed in unavailable forms in these soils. High soil pH also decreases copper availability. Mobility of Cu^{2+} in the plant is low and common deficiency symptoms include dieback in citrus and blasting of onions. Leaves of copper deficient vegetable crops lose turgor and develop a bluish-green shade before becoming chlorotic and curling. Some plants may also fail to flower, and in the case of the small grains, fail to develop heads when copper is deficient. [also see *turgor*; *chlorotic*; *plant nutrients, essential*; and *cation*]

Copper sulfate ($CuSO_4 \cdot 5H_2O$). A common source of copper for fertilizer with 25 percent copper. Copper sulfate is also used as an insecticide and fungicide, e.g., in rice fields to help control algae. The solubility of $CuSO_4 \cdot 5H_2O$ is ~23 g per 100 ml (~1.92 lbs. per gal) meaning it is fairly soluble in water.

Coprogeneous earth. A type of limnic layer (sedimentary peat) composed predominantly of fecal material derived from aquatic animals.

Cork. The outer layer of the bark in woody plants which is composed of dead cells.

Corrosion. A process of erosion where rocks and soil are removed or worn away by natural chemical processes, especially by the solvent action of running water; but also by other reactions, such as hydrolysis, hydration, carbonation, and oxidation.

Cortex. The tissue region of a stem or root surrounded externally by the epidermis and internally the vascular system. The cortex is a primary-tissue region.

Covalent (bond). A nonionic chemical bond formed by a sharing of electrons between two atoms.

Cover crop. A dense-cover crop grown when other crops are not grown (e.g., during the winter) that provides soil protection, seeding protection, and soil improvement. When plowed under or otherwise incorporated into the soil, cover crops may be referred to as "green manure crops."

Creep, soil. Slow mass movement of soil and soil material down slopes driven primarily by gravity, but facilitated by saturation with water, by alternate freezing and thawing, and by domestic and wild animals (e.g., deer and/or cattle).

Crop nutrient requirement. The amount of essential plant nutrient, or nutrients, needed to grow a specified yield of a crop plant per unit area.

Crop rotation. The practice of growing different drops in regular succession to aid in the control of insects and diseases, to increase soil fertility, and to decrease erosion.

Crotovina (also, krotovina). (1) A former animal channel in one soil horizon that has been filled with organic or inorganic soil material from another soil horizon. (2) A passageway created by an animal that becomes backfilled with soil is known as a crotovina. Also associated with earthworm tunnels or passageways.

Crown. The junction of the root and stem, usually at the level of the ground.

Crust. A surface layer on soils, ranging in thickness from a few millimeters to perhaps as much as 25 millimeters (approximately 1 inch) that is much more compact, hard, and brittle when dry than the material immediately beneath it.

Crust (Earth's). The outermost layer of the Earth, varying in thickness from 10 kilometers (about 6.2 miles) below the oceans, to 65 kilometers (about 40.4 miles) below the continents; represents less than 1 percent of the Earth's volume.

Crystal. A regular arrangement of atoms in space.

Crystal structure. The orderly arrangement of atoms in a crystalline material.

Crystallization. Physical or chemical process or action which results in the formation of regularly-shaped, -sized, and -patterned solid forms known as crystals.

Cubic feet per second (cfs). A rate of the flow, in streams and rivers, for example. It is equal to a volume of water one foot high and one foot wide flowing a distance of one foot in one second. One cubic foot per second is equal to 7.4805 gallons of water flowing each second.

Cultivar. A cultivar is a variety of plants found only under cultivation.

Cumulic layer. A cumulic layer is a layer or layers of mineral material in organic soils. Either the combined thickness of the mineral layers is more than 5 centimeters or a single mineral layer 5 to 30 centimeters thick occurs. One continuous mineral layer more than 30 centimeters thick in the middle or bottom tier is a ferric layer.

Cupric (Cu^{2+}), cupric ion. (1) The copper (II) ion, Cu^{2+}. (2) A compound that contains copper in the 2^+ oxidation state. [also see *plant nutrients, essential*; and *cation*]

Cutans. Cutans are coatings or deposits of material on the surface of peds, stones, etc. A common type of cutan is the clay cutan caused by translocation and deposition of clay particles on ped surfaces.

Cuticle. (1) A waxy layer that seals the outer surface of land plants, helping to retain moisture. (2) A film composed of wax and cutin that occurs on the external surface of plant stems and leaves and helps to prevent water loss.

Cyanamid. See *calcium cyanamid ($CaCN_2$)*.

Cyanobacteria. (1) The accurate name for N_2-fixing microbes; previously called blue-green algae. (2) Blue-green bacteria; unicellular or filamentous chains of cells that carry out photosynthesis. [also see *algae*]

Cyst. (1) The resting stage formed by some bacteria, nematodes, and protozoa in which the whole cell is surrounded by a protective layer. (2) A small, capsule-like sac that encloses an organism in its resting or larval stage, e.g., a resting spore of an alga. A cyst is not the same as an endospore.

Cytoplasm. The cytoplasm is all the contents of a cell, including the plasma membrane, but not including the nucleus. (2) The viscous semi-liquid inside the plasma membrane of a cell; contains various macromolecules and organelles in solution and suspension. [also see *protoplasm*]

Damping off. The term used for a number of different fungus-caused diseases which can kill seeds or seedlings before or after they germinate especially if in warm, wet conditions which speed growth but are conducive to fungal attacks. A given seed can become infected with a fungus often causing it to darken and soften. This will generally kill it before the seedling emerges, or cause the seedling to be weak, sometimes getting "wet" patches on it which decay until it falls apart. A seedling can be infected after it sprouts, before it leaves the ground, or even after it appears well-developed, the latter often resulting in the plant inexplicably thinning where it touches the ground, until its stem at that point rots and it falls over.

Dark reactions. The photosynthetic process in which food (sugar) molecules are formed from carbon dioxide from the atmosphere with the use of ATP. The process can occur in the dark as long as ATP is present.

Debris. Any surface accumulation of loose material detached from rock masses by chemical and mechanical means, as by decay and disintegration. Debris consists of rock clastic material of any size, and is sometimes organic matter.

Deciduous. Means falling off at maturity or tending to fall off, and is typically used in reference to trees or shrubs that lose their leaves seasonally; and to the shedding of other plant structures such as petals after flowering or fruit when ripe. In a more specific sense, deciduous means the dropping of a part that is no longer needed, or falling away after its purpose is finished. In plants it is the result of natural processes. In agriculture and horticulture, deciduous plants, including trees, shrubs and herbaceous perennials, are those that lose all of their leaves for part of the year. This process is called abscission. In some cases leaf loss coincides with winter; primarily in temperate or polar climates. [also see *abscission*]

Decomposers. (1) Organisms (bacteria, fungi, heterotrophic protists) in an ecosystem that break down organic material (dead animals and plants) into smaller molecules that are then recirculated. (2) Heterotrophic organisms that break down organic compounds, and in the process convert their organic nitrogen (found in proteins and nucleic acids) into inorganic ammonium (NH_4^+).

Decomposition (of plants and/or animals). Decomposition is the breaking down of organic materials into smaller molecules which are then recycled into nature. To break or separate into basic components or parts. Decomposition reactions are catalyzed by enzymes. [Enzyme decomposition example: cellulase. Cellulase breaks celluloses (cell wall fiber, wood) which are chains hundreds of sugar units long, into individual component sugars. Decomposition is important in organic matter decay].

Deep banding. Fertilizers and soil amendments placed in strips at a depth of 14 to 18 centimeters (5.5 to 7 inches or deeper); placed at planting or in preplant with regard to rows.

Deficiency, nutrient. The total lack or insufficient presence of essential plant nutrients in soils or in plant tissues. Deficiencies generally can be determined by means of soil and/or plant tissue analyses.

Deficiency symptom. A consequence or effect caused by the lack of an essential plant nutrient. Deficiency symptoms include slow plant growth, chlorosis, and/or necrosis.

Deflation. Preferential removal of fine soil particles from the surface of the soil by wind.

Deflocculation, soil. The opposite of flocculation. When soil solutions are low in ionic strength and dominated by alkali metal cations (especially sodium), often at higher pH values, soil colloidal particles can be dispersed throughout the solution. Deflocculation is augmented by high exchangeable soil sodium and magnesium in relation to calcium. Note: Calcium, in the form of calcium sulfate [i.e., anhydrite ($CaSO_4$) and/or gypsum ($CaSO_4 \cdot 2H_2O$)] is added to soils to help prevent soil deflocculation. <u>The preferred calcium to (sodium + magnesium) ratio in soils is 8:1; in irrigation water the preferred calcium to (sodium + magnesium) ratio is 2:1</u>] [also see *dispersion;* and *flocculation*]

Degradation. (1) The process whereby a compound is usually transformed into simpler compounds. (2) The wearing down or away, and the general lowering of the land surface by natural processes of weathering and erosion (e.g., the deepening by a stream of its channel) and may infer the process of transportation of sediment.

Degraded soil. Soil that is less productive than previously, usually because of damages from erosion, loss of humus, loss of fertility, or accumulated detrimental salts (often due to low soil calcium) or other pollutants.

Delta. A body of alluvium, nearly flat and fan-shaped, deposited at or near the mouth of a river or stream where it enters a body of relatively quiet water, usually a sea or lake.

Denitrification. The biological process by free-living soil bacteria where nitrate (NO_3^-) or nitrite (NO_2^-) is reduced and lost to gaseous nitrogen [either molecular nitrogen (N_2) or as an oxide of nitrogen]. Denitrification is augmented by poor aeration in the soil (e.g., soils with excessive amounts of water, large amounts of oxidizable carbon sources, and low amounts of soil calcium). [also see *nitrification*]

Density. (1) The mass per unit volume of a substance or solution. (2) The weight per volume at a specific temperature.

Density of Some Common Materials				
Material	kg m^{-3}	Mg m^{-3}	g cm^{-3}	lbs ft^{-3}
water	1000	1	1	62.44
pine wood	700	0.7	0.7	44
loose sand	1600	1.6	1.6	100
quartz mineral	2600	2.6	2.6	162
steel	7700	7.7	7.7	480
lead	11,300	11.3	11.3	706

Miller and Gardiner. Soils in Our Environment. 11th ed. 2007.

[Example: water weighs 8.3453 pounds per gallon. This is one way in which density is expressed]. [also see *bulk density;* and *particle density*]

Denudation. (1) Sculpturing of the surface of the land by weathering and/or erosion. (2) Leveling mountains and hills to flat or gently undulating plains.

Depletion. The removal of essential plant nutrients from the soil by leaching of water and by plant growth (particularly those plants or crops that are harvested).

Deposit. Earth material of any type, either consolidated or unconsolidated, that has accumulated by natural processes.

Deposition. (1) The laying down of any material by any agent such as wind, water, ice or by other natural processes. (2) Any accumulation of material, by mechanical settling from water or air, chemical precipitation, evaporation from solution, etc.

Desalinization (also desalination). (1) Removal of salts from saline soil, usually by leaching. (2) The removal of salts from brackish (slightly salty) water or seawater to provide freshwater that is fit to drink (potable). This method is becoming a more popular way of providing freshwater to populations.

Desert crust. A desert crust is a hard surface layer in desert regions containing calcium carbonate, calcium sulfate, or other cementing materials.

Desert pavement. Desert pavement is a layer of gravel or stones remaining on the surface of the ground in deserts after the removal of fine material by wind action.

Desert varnish. Also called rock varnish, a dark coating found on exposed rock surfaces in arid environments. The desert varnish forms only on physically stable rock surfaces that are no longer subject to frequent precipitation, fracturing or wind abrasion. The varnish is primarily composed of particles of clay along with iron and manganese oxides. There is also many trace elements and almost always some organic matter. The color of the varnish varies from shades of brown to black. Originally scientists thought that the varnish was made from substances drawn out of the rocks it coats. However, microscopic and chemical observations/tests show that a major part of varnish is clay which could only arrive by wind. The clay then acts as a substrate to catch additional substances that chemically react together when the rock reaches high temperatures in the desert sun. Wetting by dew is also important in the process.

Desertification. The degradation of land in arid and dry sub-humid areas due to various factors including climatic variations and human activities. A major impact of desertification is reduced biodiversity and diminished productive capacity.

Desorption. Desorption is the removal of adsorbed entities off of the adsorption sites, and the inverse of adsorption. [also see *sorption*; *adsorption;* and *absorption*]

Detritus. (1) Dissolved and particulate dead organic matter, or any fragmentary material or disintegrated matter. (2) Carbon rich organic material decomposing at varying rates depending on factors such as temperature and soil moisture conditions. The active fraction of soil organic matter, as opposed to humus, the resistant fraction. [also see *detrivore*; *organic matter, soil;* and *humus*]

Detrivore. Any organism that obtains most of its nutrients from the detritus (non-living particulate organic material; typically including the bodies or fragments of dead organisms as well as fecal material) in an ecosystem.

Deoxyribonucleic acid (DNA). A polymer of nucleotides connected via a phosphate-deoxyribose sugar backbone; the genetic material of the cell.

Diammonium phosphate (DAP) $(NH_4)_2HPO_4$ (18-46-0). A fertilizer composed of ammonium phosphates, principally diammonium phosphate, resulting from the ammoniation (to treat or combine with ammonia) of phosphoric acid. It may contain up to 2% nonammoniacal nitrogen. The guaranteed percentage of nitrogen and available phosphate should be stated as part of the name.

Diatomaceous earth. (1) Geologically deposited fine mineral material of fine, grayish, siliceous material composed primarily, or wholly, of the remains of skeletons of algae (diatoms) or other nanoplankton. It may occur as a power or as a porous, rigid material. (2) Fossilized deposits of diatoms used for abrasives, polishes and as a filtering agent.

Diatomite. Diatomaceous earth.

Diatoms. Diatoms are algae that possess a siliceous cell wall that remains preserved after the death of the organisms. Diatoms are abundant in both fresh and salt water and in a variety of soils.

Diazotroph. An organism that can use dinitrogen (N_2) as its sole nitrogen source, i.e. capable of N_2 fixation.

Dicalcium phosphate $(CaHPO_4 \cdot 2H_2O)$. A manufactured fertilizer consisting chiefly of dicalcic salt of phosphoric acid.

Dicotyledons (also known as dicots). One of the two main types of flowering plants; characterized by having two cotyledons, floral organs arranged in cycles of four or five, and leaves with reticulate veins; includes trees (except conifers) and most ornamental and crop plants. [also see *monocotyledons*]

Diffusion (especially, nutrient). (2) The net movement of ions in water from a more concentrated region to a less concentrated region, mostly by its own kinetic motion or a chemical activity gradient. (2) Movement of nutrients in soil that results from a concentration gradient. [also see *mass flow;* and *root interception*]

Digester. An enclosed composting system with a device to help mix and aerate the waste materials.

Digestion (composting). The most active stage of the composting process which is usually carried out in open windrows or in enclosures. The objective of compost digestion is to create an environment in which microorganisms will rapidly decompose the organic portion of the composting material.

Dihydrate. Refers to the hydrated calcium sulfate compound ($CaSO_4 \cdot 2H_2O$) also known as gypsum. [also see *gypsum;* and *anhydrite*]

Dinitrogen fixation (also biological nitrogen fixation BNF). Conversion of gaseous molecular nitrogen (N_2), by certain soil bacteria, algae, and actinomycetes, to ammonia (NH_3). The ammonia is subsequently converted to organic nitrogen utilizable by plants and animals. The nitrogen-fixing process associated with legume nodules-roots is known as symbiotic nitrogen fixation, while nitrogen fixation by soil organisms not associated with higher plants is termed non-symbiotic. [also see *nitrogenase*]

Biological nitrogen fixation occurs when atmospheric nitrogen is converted to ammonia by an enzyme called nitrogenase. The formula for nitrogen fixation is:

$$N_2 + 8H^+ + 6e^- \longrightarrow 2NH_3 \text{ (ammonia)} + H_2$$

Disaccharide. Any of a class of sugars, such as maltose, lactose, and sucrose, having two linked monosaccharide units per molecule. [also see *monosaccharide;* and *polysaccharide*]

Discharge. The volume of water that passes a given location within a given period of time. Usually expressed in cubic feet per second.

Disease. Organisms suffer from disease when their normal function is impaired by some genetic disorder, or more often from the activity of a parasite or other organism living within them. Viruses, bacteria, and fungi cause many diseases.

Dispersion. The breaking down of soil aggregates into individual particles. Generally, the more easily dispersed the soil, the more erodible it is; also leading to decreased air movement, water infiltration and root growth. Dispersion is augmented by high exchangeable soil sodium and magnesium in relation to soil calcium. [also see *deflocculation*]

Diversity. The different types of organisms that occur in a community.

Dolomite. A carbonate sedimentary rock composed of more than 50% of the mineral calcium magnesium carbonate [$CaMg(CO_3)_2$]. A mineral used for liming soils. [also see *dolomitic lime;* and *limestone*]

Dolomitic lime ($CaCO_3 \cdot MgCO_3$). A natural occurring liming material composed chiefly of carbonates of Mg and Ca in approximately equimolar proportions. Dolomitic lime contains approximately 110% $CaCO_3$ equivalent, and 36% Ca plus 20% Mg, depending upon the source. [also see *limestone*]

Duripan. A sub-surface horizon that is cemented mostly by silica. In California, duripans are commonly associated with the San Joaquin soil series (i.e., the official California State Soil) found in the foothills of the Sierra Nevada mountains and eastern San Joaquin and Sacramento valleys. [also see *pan; cemented (soil); hardpan;* and *claypan*]

Drainage, excessive. Too much or too rapid loss of water from soils, either by percolation or by surface flow. The occurrence of internal free water is very rare or very deep.

Drainage, surface. Used to refer to surface movement of excess water; includes such terms as ponded, flooded, slow, and rapid.

Drawdown. The lowering of the ground-water surface caused by pumping.

Drip irrigation. See *irrigation, drip*.

Dry weight, soil. The equilibrium weight of the solid soil particles after the water has been vaporized by heating to 105 °C (221°F).

Dryland farming. (1) Rain fed farming in arid and semiarid regions without the use of irrigation. (2) Farming on non-irrigated land. Success is based on rainfall, moisture-conserving tillage, and drought-resistant crops.

Dunes, sand. Ridges or small hills of sand, which have been piled up by wind action on sea coasts, in deserts, and elsewhere.

Dynamic pile system (composting). Compost piles receive forced aeration and are not turned or mixed.

"E" horizon, soil. A soil horizon characterized by maximum illuviation (washing out) of silicate clays and iron and aluminum oxides. Commonly occurs above the B horizon and below the A horizon. [also see *illuviation;* and *eluviaton*]

Ecology. The branch of biology that focuses on the relationships among various species and among the species and their environments. The basic units for study are the species (all the organisms which are capable of interbreeding), population (all of the members of a species occupying a certain geographical area) and community (number of populations interacting within a certain area). The emphasis in environmental science, however, is on the processes of energy flows and nutrient cycles within these communities and ecosystems; and involves such ideas as species abundance, biodiversity, complexity, trophic (feeding status) levels, ecological niches, and structure of communities.

Ecosystem. (1) A major interacting system that involves both living organisms and their physical environment. (2) The community living in an area and its physical environment.

Ectomycorrhiza. A mycorrhizal association in which the fungal mycelia extend inward, between root cortical cells, to form a network and outward into the surrounding soil. Usually the fungal hyphae also form a a mantle on the surface of the roots. [also see *endomycorrhiza; mycorrhiza; symbiosis;* and *vesicular arbuscular*]

Edaphic. (1) Pertaining to the soil. (2) Influenced by soil factors.

Edaphology. The science that deals with the influence of soils on living things, particularly plants, including man's use of land for plant growth.

Efflorescence. The accumulation of dissolved substances (usually simple salts) at the soil surface due to evaporation.

Effluent. The discharge or outflow of water from ground or subsurface storage. The fluids discharged from domestic, industrial, and/or municipal waste collection systems or treatment facilities.

Effusion. The movement of gas molecules through an opening that has relatively large holes.

Electrical conductivity (EC). Conductivity of electricity through water and/or soil extracts. A measure of total salt concentration of soils and irrigation water, expressed in terms of $dS\ m^{-1}$ (decisiemens per meter) or millimhos per centimeter (outdated, but the numbers = $dS\ m^{-1}$). Soils: EC measurements less than $0.60\ dS\ m^{-1}$ in soils are considered destructively low; and may inhibit irrigation water penetration from surface soil sealing due to the lack of sufficient salt content. EC measurements above $4.00\ dS\ m^{-1}$ indicate salinity hazards due to excessive total salt content in soil. Irrigation water: EC measurements less than $0.60\ dS\ m^{-1}$ are low and may contribute to poor irrigation water penetration. EC measurements greater than $3.0\ dS\ m^{-1}$ are severely saline and likewise contribute to soil salinity hazards and the accumulation/buildup of harmful salts in soil.

Electrical conductivity problems in soils and irrigation water are commonly corrected/treated with the use of calcium sulfate products [i.e., anhydrite ($CaSO_4$) and gypsum ($CaSO_4 \cdot 2H_2O$)]. [also see *saline soil*]

Electrolyte. A substance that dissociates into ions in aqueous solution and so makes possible the conduction of an electric current through the solution. Very pure water is a poor conductor of electricity. However, very small amounts of impurities, especially salts, greatly increase the electrical conductivity of water. [Example: NaCl (table salt)]

Electron. A subatomic particle with a negative charge. Electrons circle the atom's nucleus in regions of space known as orbitals.

Element. (1) A substance composed of only one kind of atom that cannot be decomposed by a chemical change into simpler substances. (2) The simplest form of matter each of which has unique physical and chemical properties. Elements, singly or in combination, compose virtually all materials of the universe.

It is frequently quoted that there are "92 naturally occurring chemical elements," but this is incorrect. <u>There are actually only 88 naturally occurring chemical elements</u>. The elements technetium (43), promethium (61), astatine (85) and francium (87) have no stable isotopes, and none of long half-life, so they are not naturally present or occurring. [also see *atom; periodic table of the elements;* and *plant nutrients, essential*]

Element, essential. See *plant nutrients, essential.*

Elemental composition (of plants). See *plant nutrients, essential.*

Eluvial soil horizon. A soil horizon from which material (e.g., clays and/or humus) has been removed either in solution or suspension.

Eluviation. The downward removal of soil material in suspension (or in solution) from a layer of soil. [also see *illuviation*]

Endomycorrhiza. A mycorrhizal association with intracellular penetration of the host root cortical cells by the fungus as well as outward extension into the surrounding soil. [also see *ectomycorrhiza; mycorrhiza; symbiosis;* and *vesicular arbuscular*]

Endophyte. An organism growing within a plant. The association may be symbiotic or parasitic.

Endosymbiosis. When one organism takes up permanent residence within another, such that the two become a single functional organism. Mitochondria and plastids are believed to have resulted from endosymbiosis.

Entomology. The study of insects and their environments.

Environmental sustainability. Environmental sustainability implies the following:

- meeting the basic needs of all peoples, and giving this priority over meeting the greed of a few
- keeping population densities, if possible, below the carrying capacity of the region
- adjusting consumption patterns and the design and management of systems to permit the renewal of renewable resources
- conserving, recycling, and establishing priorities for the use of nonrenewable resources
- keeping environmental impact below the level required to allow the systems affected to recover and continue to evolve

An environmentally sustainable agriculture is one that is compatible with and supportive of the above criteria. [also see *sustainable agriculture*]

Enzyme. (1) A protein that is capable of speeding up specific chemical reactions by lowering the required activation energy, but is unaltered itself in the process. (2) Protein within or derived from a living organism that functions as a catalyst to promote specific reactions. Decomposition reactions are catalyzed by enzymes. Each enzyme is given a name descriptive of the particular reaction its does, plus the ending "–ase." [Example: the enzyme urease breaks down urea [$CO(NH_2)_2$] to water, carbon dioxide, and ammonia (NH_3)]

Eolian (soil material). Soil material accumulated through wind action. The extensive areas in the United States are silty deposits (*loess*), but large areas of sandy deposits also occur. [also see *aeolian;* and *loess*]

Ephemeral stream. A stream that flows only intermittently, such as after storms.

Epidermis (epidermal cells). The outermost layer of cells of the leaf and of young stems and roots. This tissue often contains specialized cells for defense, gas exchange, or secretion. [also see *guard cells*]

Epipedon. Greek *epi* (over, upon), *and pedon* (soil). A soil horizon that forms at or near the surface, and in which most of the rock structure has been destroyed. It is darkened by organic matter or shows evidence of eluviation, or both. [also see *eluviation*]

Summary of Epipedon Names and Important Characteristics

Epipedon Name	Derivation	Important Characteristics
Plaggen	*plaggen*, sod (German)	Overly thick mollic (> 50 cm) due to long continued manure application
Histic	*histos*, tissue (Greek)	Thin, organic horizon saturated 30 consecutive days or more, unless drained. If mixed with mineral material, remains very high in organic matter
Anthropic	*anthropos*, man (Greek)	Like mollic, but with a high phosphorus content due to long period of cultivation and fertilization
Mollic	*mollis*, soft (Latin)	Thick, well-structured, base saturation > 50 %, dark-colored mineral soil horizon
Umbric	*umbra*, shade (Latin)	Like mollic, but with base saturation < 50 %
Ochric	*ochros*, pale (Greek)	Surface mineral horizon that does not meet criteria for other epipedons

Compiled From Various Sources

Epsom salt. See *magnesium sulfate*.

Equivalent (eq): An equivalent, in a neutralization reaction, is the mass of acid that yields 1 mole of H^+, or the mass (in grams) of base that reacts with 1 mol (1 g) of H^+. Example: is the following neutralization reaction:

$$H_2SO_4 \text{ (sulfuric acid)} + 2\ NaOH \text{ (lye)} \longrightarrow Na_2SO_4 + 2H_2O$$

In the above reaction, each mole of H_2SO_4 supplies two moles H^+. Because the molar mass is 98.1 for H_2SO_4, an equivalent in this reaction equals 98.1 grams ÷ 2 = <u>49.9 g</u>. Because each mole of NaOH reacts with 1 mole H^+, and equivalent of NaOH equals its molar mass <u>40.0 g</u>. One equivalent of H_2SO_4 (49.9 g) reacts with one equivalent of NaOH (40.0 g). [also see *mole;* and *milliequivalents per liter*]

Equivalent Weights (g) of Common Soil Ions and Other Chemicals

Name	Symbol	Atomic Wt.	Equivalent Wt.
Calcium	Ca^{2+}	40.08	20.04
Magnesium	Mg^{2+}	24.31	12.15
Sodium	Na^+	22.99	22.99
Potassium	K^+	39.10	39.10
Bicarbonate	HCO_3^-	61.02	61.02
Carbonate	CO_3^{2-}	60.01	30.00
Sulfate	SO_4^{2-}	96.06	48.03
Chloride	Cl^-	35.46	35.46
Nitrate	NO_3^-	62.01	62.01
Ammonium	NH_4^+	17.03	17.03
Calcium chloride	$CaCl_2$	110.98	55.49
Anhydrite	$CaSO_4$	136.14	68.07
Gypsum	$CaSO_4 \cdot 2H_2O$	172.18	86.09
Calcium carbonate	$CaCO_3$	100.08	50.04
Dolomitic lime	$CaCO_3 \cdot MgCO_3$	184.41	92.20
Magnesium sulfate	$MgSO_4$	120.38	60.19
Magnesium carbonate	$MgCO_3$	84.32	42.16
Sodium chloride	$NaCl$	58.46	58.46
Potassium chloride	KCl	74.56	74.56
Potassium sulfate	K_2SO_4	174.26	87.13
Sulfur	S	32.07	16.03

Compiled From Various Sources

Erosion. The wearing away of the land surface by rain or irrigation water, wind, ice, or other natural or anthropogenic agents that abrade, detach and remove geologic parent material or soil from one point on the Earth's surface and deposit it elsewhere. Erosion includes such processes as gravitational creep and tillage erosion.

The following terms are used to describe different types of water erosion:

Accelerated erosion. Erosion much more rapid than normal, natural, geological erosion; primarily as a result of the activities of human or, in some cases, animals.

Gully erosion. The erosion process whereby water accumulates in narrow channels and, over short periods, removes the soil from this narrow area to considerable depths, ranging from 1 to 2 feet to as much as 75 to 100 feet.

Natural (geological) erosion. Wearing away of the Earth's surface by water, ice, or other natural agents under natural environmental conditions of climate, vegetation, etc., undisturbed by man.

Rill erosion. An erosion process in which numerous small channels of only several inches (or centimeters) in depth are formed. Rill erosion occurs mainly on recently cultivated soils.

Sheet erosion. The removal of a fairly uniform layer of soil from the land surface by runoff water.

Splash erosion. The spattering of small soil particles caused by the impact of raindrops on very wet soils. The loosened and separated particles may or may not be subsequently removed by surface runoff.

Erosion pavement. A layer of gravel or stones left on the surface of the soil or ground after the removal of the fine particles by erosion.

Escarpment. A relatively continuous cliff or relatively step slope, produced by erosion or faulting, breaking the general continuity of more gently sloping land surfaces. The term is most commonly applied to cliffs produced by differential erosion and it is commonly used synonymously with *scarp*. [also see *scarp*]

Essential plant nutrients (elements). See *plant nutrients, essential*.

Estuary. (1) A seaward end or the widened funnel-shaped tidal mouth of a river valley where fresh water comes into contact with seawater and where tidal effects are evident. (2) A tidal river, or a partially enclosed coastal body of water where the tide meets the current of a stream.

Eukaryote. All species of large complex organisms are eukaryotes, including animals, plants and fungi, although most species of eukaryotic protists are microorganisms. [also see *protista*]

Eutectic (solutions). When salts are added to water they depress the freezing point of the water (Example: this is why salting roads and sidewalks melts snow on them). Adding more salt generally depresses the freezing temperature further, but these solutions do not freeze cleanly and at a precise temperature, instead they tend to form a slush. However, if a particular salt at a particular concentration is added, the resulting solution freezes and melts cleanly at a constant temperature, releasing and storing large amounts of energy as it does so. This temperature is called the eutectic point and the composition is called a <u>eutectic solution.</u>

Eutrophic. (1) Pertaining to or aiding nutrition. (2) Any agent that promotes nutrition. (3) Having high concentrations of nutrients optimal, or nearly so, for plant or animal growth. Can be applied to nutrient or soil solutions and bodies of water.

Eutrophication. Nutrient enrichment of water. Eutrophication is usually most affected by phosphorus and less so by nitrogen. Increased algal growth eventually loads the water with dead algae, which, during microbial decomposition, results in consumption of the water's dissolved oxygen (especially during the warm summer months), causing water life to die (or is detrimental to other organisms). [also see *algal blooms*]

Evaporation. (1) The part of the hydrologic cycle in which liquid water is converted to vapor and enters the atmosphere. (2) The process of the change in the state of a liquid or solid to a gas or vapor. Vanishing of the surface of a liquid to the atmosphere. Evaporation is the opposite of condensation. [also see *condensation*]

Evaporite. A deposit of salt minerals (e.g., halite, gypsum, anhydrite) left behind by the evaporation of seawater. Evaporites usually form within a restricted basin.

Evapotranspiration (ET). The total water loss due to the transpiration of vegetation plus the evaporation from the soil; higher climatic temperatures result in a higher evapotranspiration rates.

Evergreen. Plants, shrubs, and trees that hold their leaves or needles for a long period of time. Evergreens do not display pronounced outward changes such as deciduous trees do when they enter dormant periods. Most deciduous plants drop their leaves by abscission before winter, while evergreen plants continuously abscise their leaves. [also see *deciduous;* and *abscission*]

Excessively aerobic. A soil horizon which is usually too dry to support adequate plant growth.

Excessively drained. A soil that loses water very rapidly because of rapid percolation.

Exchangeable acidity. See *acid soil (soil acidity.*

Exchangeable ion. (1) Any ion held through electrical attraction to a charged surface. Ions can be displaced by other ions of like charge from the surrounding solution. (2) Exchangeable ions are loosely defined as those released from the soil by solutions of neutral salts [e.g., anhydrite ($CaSO_4$) or gypsum ($CaSO_4 \cdot 2H_2O$)]. (3) Exchangeable cations are readily available for plants. The sum of exchangeable calcium, magnesium, potassium, sodium, and aluminum generally equals, for all practical purposes, the soil's cation exchange capacity. [also see *soluble salts;* and *cation exchange*]

CEC Values and Major Exchangeable Cations Average of Agricultural Soils (California)						
pH	CEC (mmole kg^{-1})	Exchangeable Cations (% of Total)				
		Ca^{2+}	Mg^{2+}	K^+	Na^+	H^+ and Al^{3+}
7.0	203	65.6	26.3	5.5	2.6	---

Bohn, et. al. Soil Chemistry, 2nd ed. 2001.

Important Note: Exchangeable H^+ and Al^{3+} are only appreciably present in strongly acidic soils (pH < 4)

Exchangeable sodium percentage (ESP). The percentage of the soil exchange complex saturated or occupied by sodium. Exchangeable sodium contents can affect permeability and indicate sodium toxicity hazards. ESP may be calculated by the formula:

$$ESP = \frac{\text{exchangeable sodium (meq/100 grams soil)}}{\text{cation exchange capacity (meq/100 grams soil)}}$$

Exchangeable Sodium Percentages	
Below 5	Generally no soil permeability problems are expected. Some sodium sensitive crops, perennial trees and vine crops may show phytotoxicity symptoms
10	Technically determined to be a sodic soil
10-15	Possible water permeability problems with finer textured soil (saturation percentages greater than 50% [clayey soils])
Above 15	Permeability problems occur on all mineral soils except those having saturation percentages less than 20% [sandy soils]. Phytotoxicity is likely for most crops

Miller and Gardiner. Soils in Our Environment. 11th ed. 2007.

Excretion. The process of removing the waste products of cellular metabolism from the body.

Exfoliation. A weathering process where thin layers of rock peel off from the surface. This is caused by the heating of the rock surface during the day and cooling at night leading to alternate expansion and contraction of the rock surface. This process is sometimes termed "onion skin weathering."

Exponential growth. A period of sustained growth of a microorganism in which the cell number constantly doubles within a fixed time period.

Extracellular. Located or occurring outside a cell or cells. [Example: extracellular fluid]

Extract, soil. The solution separated from a soil suspension or from a soil by filtration, centrifugation, suction, or pressure. (May or may not be heated prior to separation).

Extractable (nutrients). The term "extractable" is often used and is synonymous with available and exchangeable. It refers to the amount of nutrients that can be extracted from the soil solution and is ready for plant uptake.

Extrusive. Igneous. [also see *intrusive*]

Exudates, root. Low molecular weight metabolites that enter the soil after being excreted or secreted from plant roots. These include at least 18 amino acids, 10 sugars, 10 organic acids, and various proteins, growth substances, growth inhibitors, microbe attractants, and repellents.

Facultative organisms. (1) Organisms capable of growing under both aerobic and anaerobic conditions. (2) Organisms that can carry out both options of a mutually exclusive process (e.g., aerobic and anaerobic metabolism).

Facultative symbiont. Symbiotic relationships may be either obligate, i.e., necessary for the survival of at least one of the organisms involved, or facultative, where the relationship is beneficial but not essential for survival of the organisms. [also see *obligate symbiont;* and *symbiont*]

Fallow. The practice of leaving land either uncropped and weed-free, or with volunteer vegetation during at least one period when a crop would normally be grown. The objectives may be to: (a) control weeds, (b) accumulate water, and/or (c) accumulate available plant nutrients.

Fauna. (1) Term referring collectively to all animals in an area. The zoological counterpart of flora. (2) Animals or animal life of a particular region or a particular time. [also see *flora*]

Fecal matter (material). The various types of feces or excrement produced by soil fauna.

Feedstock. The raw material used for chemical or biological processes. For example, in composting, feedstock might include grass clippings, leaves, food scraps, plant trimmings, straw, and/or animal bedding.

Fen. Low, flat, swampy land; a bog or marsh.

Fen peat. Organic peat that is neutral to alkaline due to the presence of calcium carbonate.

Fermentation. (1) The extraction of energy from organic compounds without the involvement of oxygen. (2) Metabolic processes in which organic compounds serve as both electron donors and electron acceptors.

Ferric, (ferric ion). (1) The oxidized iron (III) ion, Fe^{3+}. (2) A ferric compound is one that contains iron in the +3 oxidation state. At pH = 3, ferric iron is soluble enough to supply plant needs. However, the solubility of Fe^{3+} decreases about a thousand fold per pH unit rise. Therefore, most of the soluble iron in the soil is not found in this form but in the ferrous (Fe^{2+}) form.

Ferrous, (ferrous ion). (1) The reduced iron (II) ion, Fe^{2+}. (2) A ferrous compound is one that contains iron in the +2 oxidation state. The reduced ionic form of iron, formed in anaerobic conditions, is more soluble than the ferric ion, but ferrous iron is also easily and rapidly oxidized in aerated soils. However, most of the soluble iron found in soils is in this form.

Ferrous sulfate ($FeSO_4 \cdot 7H_2O$). See *iron sulfate*.

Fertigation. Application of plant nutrients and soil amendments through irrigation water. [also see *chemigation*]

Fertility, residual. (1) The available nutrient content of a soil carried over to subsequent crops or years. (2) Available nutrient content of a soil carried over to the next crop after fertilizing the previous crop. [also see *residual value*]

Fertility, soil. (1) The relative quality of a soil that enables it to supply/provide nutrients essential to plant growth. (2) The quality of a soil that enables it to provide nutrients in adequate amounts and in proper balance for the growth of specified plants or crops, when other factors such as light, moisture, temperature, and physical condition of the soil are positive or favorable. The importance of soil fertility and plant nutrition to the health and survival of all life cannot be overstated. However, a fertile soil is not necessarily a productive soil unless other limiting factors are also controlled. Adequate amounts of plant nutrients are but one characteristic of a productive soil.

Fertilizer. Any organic or inorganic material of natural or synthetic origin (other than liming materials) that is added to a soil to supply one or more plant nutrients essential to the growth of plants. Any substance containing one or more recognized plant nutrient(s) which is used for its plant nutrient content and which is designed for use of claimed to have value in promoting plant growth, except un-manipulated animal and vegetable manures, marl, lime, limestone, wood ashes and other exempt products.

Common Name	Formula	N	P_2O_5	K_2O
Nitrogen Materials				
Ammonium nitrate	NH_4NO_3	34	0	0
Ammonium sulfate	$(NH_4)_2SO_4$	21	0	0
Ammonium nitrate-urea	$NH_4NO_3 + CO(NH_2)_2$	32[1]	0	0
Anhydrous ammonia	NH_3	82	0	0
Aqua ammonia	NH_4OH	20[1]	0	0
Urea	$CO(NH_2)_2$	46	0	0
Ammonium nitrate	NH_4NO_3	34	0	0
Ammonium sulfate	$(NH_4)_2SO_4$	21	0	0
Calcium Cyanamid	$CaCN_2$	21		
Phosphorus Materials				
Superphosphate	$Ca(H_2PO_4)_2$	0	20	0
Concentrated superphosphate	$Ca(H_2PO_4)_2$	1	44 – 46	0
Ammoniated superphosphate	$Ca(NH_4H_2PO_4)_2$	5[1]	40[1]	0
Ammonium phosphate	$(NH_4)_3PO_4$	16	20	
Monoammonium phosphate	$NH_4H_2PO_4$	13[1]	52[1]	0
Diammonium phosphate	$(NH_4)_2HPO_4$	18[1]	46[1]	0
Urea-ammonium phosphate	$(NH_2)_2CO + (NH_4)_2HPO_4$	28	28	0
Potassium Materials				
Muriate of potash	KCl	0	0	60
Monopotassium phosphate	KH_2PO_4	0	50[1]	40
Potassium sulfate	K_2SO_4			48
Sulfate of potash-magnesia	$K_2SO_4MgSO_4$	0	0	22
Miscellaneous			(%)	
Borate	Na_2BO_4		20[1]	B
Borax	$Na_2B_4O_7 \cdot 10H_2O$		11[1]	B
Copper sulfate	$CuSO_4 \cdot 5H_2O$		25[1]	Cu
Disodium octaborate (Solubor)	$Na_2B_8O_{13} \cdot 4H_2O$		20.5	B

Ferrous sulfate	FeSO$_4$·7H$_2$O		20[1]	Fe
Magnesium sulfate	MgSO$_4$·7H$_2$O		16[1]	Mg
Magnesium oxide	MgO		45[1]	Mg
Manganese sulfate	MnSO$_4$·H$_2$O		26 - 28	Mn
Zinc sulfate	ZnSO$_4$·7H$_2$O		36[1]	Zn
Zinc oxide	ZnO		50[1]	Zn
Zinc chelate	Zn chelate		14[1]	Zn
Superphosphate	Ca(H$_2$PO$_4$)$_2$		14[1]	S
Calcium sulfate (anhydrite)	CaSO$_4$		25[1]	S
Calcium sulfate (gypsum)	CaSO$_4$·2H$_2$O		22[1]	
Sodium Molybdate	Na$_2$MoO$_4$·2H$_2$O		39	Mo

Compiled From Various Sources   [1] = Variable Analysis

Solubility of Fertilizers in Cold Water	
Material	**(lbs/100 gal)**
Ammonium nitrate	984
Ammonium sulfate	592
Calcium cyanamide	Decomposes
Calcium nitrate	851
Diammonium phosphate	358
Monoammonium phosphate	192
Potassium nitrate	108
Sodium nitrate	608
Superphosphate, ordinary	17
Superphosphate, triple	33
Urea	651
Ammonium molybdate	Decomposes
Borax	8
Calcium chloride	500
Copper oxide	Insoluble
Ferrous (iron) sulfate [reduced]	242
Foliarel	123 @ 68°F
Magnesium sulfate	592
Manganese sulfate	876
Sodium chloride	300
Sodium molybdate	467

Solubor	79 @ 68°F
Zinc sulfate	625

O.A. Lorenz & D.N. Maynard. Knott's Handbook For Vegetable Growers. 2nd ed. 1980.

Acid-forming fertilizer. See *Acid-forming fertilizer*.

Analysis, fertilizer. The percent composition of a fertilizer as determined in a laboratory and expressed as total nitrogen (N), available phosphoric acid (P_2O_5) equivalent, and water-soluble potash (K_2O) equivalent. [*also see guaranteed analysis (fertilizer)*]

Banding, fertilizer. A method of fertilizer application. Banding is general term that implies application which concentrate fertilizers into narrow zones that are kept intact to provide a concentrated source of nutrients. Applications may be made prior to, during, or after planting. (**Deep banding fertilization:** Preplant application of nutrients placed 2 to 6 inches or more below the soil surface. The applied nutrients may be in solid, fluid, or gaseous forms).

Bone meal fertilizer. See *Bone meal (6-12-0)*

Bulk fertilizer. A fertilizer distributed in a non-packaged form.

Commercial fertilizer. Any substance that contains 5% or more of nitrogen (N), available phosphoric acid (P_2O_5), or soluble potash (K_2O), singly or collectively, which is distributed for promoting or stimulating plant growth. "Commercial fertilizer" includes both "agricultural" and "specialty fertilizers." [also see *mineral, agricultural*]

Complete fertilizer. A chemical compound or a blend of compounds that contains significant quantities of all three primary nutrient, nitrogen (N), phosphorus (P), and potassium (K). It may also contain other plant nutrients.

Compound fertilizer. A mixed fertilizer containing at least two of the primary plant nutrients, N, P, and K, formed by intimately mixing two or more fertilizer materials or granulating them together, usually by the processes that involve chemical reactions of the materials with each other. [also see *mixed fertilizers*].

Concentrated fertilizer. Mixed fertilizers containing 30% or more of the primary plant nutrients, $N + P_2O_5 + K_2O$. [also see *mixed fertilizer*]

Controlled release fertilizer. See *slow-release fertilizer*.

Discharge, fertilizer. A release outside a containment area of fertilizer in a quantity exceeding fifty-five U.S. gallons and/or of dry bulk fertilizer in a quantity exceeding two hundred pounds.

Filler, fertilizer. A substance added to fertilizer materials to provide bulk, prevent caking or serve some purpose other than providing essential plant nutrients. [Example: anhydrite, gypsum, lime, or clay]

Fluid fertilizer. A fertilizer in fluid form, and includes solutions, emulsions, suspensions and slurries. Fluid fertilizer does not include anhydrous ammonia.

Formula, fertilizer. The quantity and grade of the stock materials used in making a fertilizer mixture. Example: 800 pounds of diammonium phosphate (18-46-0), 800 pounds of ammonium nitrate (34% N), and 400 pounds of potassium sulfate (53.8% K_2O). The 2000 pound total of this mixture would have an analysis of 20.8% N, 18.4% P_2O_5, and 10.8% K_2O.

Note: 800 X 0.46 = 368 X 100 divided by 2,000 = 18.4% P_2O_5
Note: 400 X 0.538 = 215.2 X 100 divided by 2,000 = 10.8% K_2O

Grade, fertilizer. The guaranteed minimum of available plant food, in percent, of the major plant nutrient elements contained in a fertilizer material or in a mixed fertilizer. The analysis is usually designated as total nitrogen (N), available phosphate (P_2O_5), and soluble potash (K_2O). [Example: a 10-20-10 fertilizer grade contains 10% N, 20% P_2O_5, and 10% K_2O]

Granular fertilizer. A fertilizer where ninety-five (95) percent or more of the product is retained on a series of sieves within the range of U.S. No. 4 (4.75 mm opening) to and including U.S. No. 20 (0.850 mm opening) and in which the largest particle passes through a sieve having an opening not larger than four (4) times that of the sieve which retains ninety-five (95) percent or more of the product. The desired size may be obtained by agglomerating smaller particles, crushing and screening larger particles, controlling size in crystallization processes, or prilling.

Guaranteed analysis. See *guaranteed analysis (fertilizer)*.

Inorganic fertilizer. A fertilizer material in which carbon is not an essential component of its basic chemical structure (Soil Science Society of America).

Kelp. See *kelp*.

Liquid fertilizer. (1) A fertilizer wholly or partially in solution that can be handled as a liquid, including clear liquids and liquids containing solids in suspension. (2) A fluid fertilizer in which the plant nutrients are in true solution.

Material, fertilizer. Either:
(a) contains important quantities of no more than one of the primary plant nutrients (nitrogen, phosphate, potash).
(b) has 85 percent (85%) or more of its plant nutrient content present in the form of a single chemical compound, or
(c) is derived from a plant or animal residue or by-product or natural material deposit which has been processed in such a way that its content of plant nutrients has not been materially changed except by purification and concentration. (AAPFCO).

Mixed fertilizer. A fertilizer containing any combination or mixture of fertilizer materials. Two or more fertilizer materials mixed, or granulated together into individual pellets. The term includes dry mixed powders, granulates (bulk blends), granulated mixtures, and clear liquid mixed fertilizers, suspensions, and slurries. [also see *compound fertilizer*]

Natural organic fertilizer. Materials derived from either plant or animal products containing one of more elements (other than carbon, hydrogen, and oxygen) that are essential for plant growth. These materials may be subjected to biological degradation processes under normal conditions of aging, rainfall, sun-curing, air drying, composting, rotting, enzymatic, or anaerobic/aerobic bacterial action, or any combination of these. These materials are not mixed with synthetic materials or changed in any physical or chemical manner from their initial state except by manipulations such as drying, cooking, chopping, grinding, shredding, hydrolysis, or pelleting. (official AAPFCO). [also see *synthetic fertilizer;* and *organic fertilizer*]

[Note: For complete information on certified organics and the USDA's National Organic Program (NOP), visit: www.ams.usda.gov/nop and the Organic Materials Review Institute, www.omri.org/]

Organic fertilizer. Soil Science Society of America and AAPFCO official definition: A material containing carbon and one of more plant nutrients in addition to hydrogen and/or oxygen. Organic fertilizers are typically by-products from the processing of animal or vegetable substances that contain sufficient plant nutrient to be of value as fertilizers. [also see *natural organic fertilizer*]

Pelletized fertilizer. A form, uniform in size and usually of globular shape, containing one or more nutrients produced by one of several methods including: (1) solidification of a melt while falling through a countercurrent stream of air, (2) dried layers of slurry applied to recycling particles, (3) compaction, (4) extrusion, and (5) granulation.

Pop-up fertilizer: Fertilizer placed in small amounts in direct contact with the seed.

Polymer coated fertilizer. A coated slow release fertilizer consisting of fertilizer particles coated with a polymer (plastic) resin and is a source of slowly available plant nutrients.

Postplant fertilizer (fertilization). Fertilizer applied after planting without specific references to method of application. Side-dressing and top-dressing are methods or forms of postplant fertilization.

Preplant fertilizer (fertilization). Fertilizer applied to the soil prior to planting

Ratio, fertilizer. The relative proportions of primary nutrients in a fertilizer grade divided by the highest common denominator for that grade. [Example: grades 10-6-4 and 20-12-8 have a ratio 5-3-2]

Salt-index, fertilizer. The ratio of the decrease in soil water potential of a solution containing a fertilizer compound or mixture to that produced by the same weight of $NaNO_3$ X 100.

Side-banded fertilizer. Placement of fertilizer in bands on one or both sides of the seed or seedlings. Placement of starter fertilizers is often termed side-banded when fertilizer material is placed to the side and below the seed on one or both sides of the row. The term may also refer to the placement of fertilizers in a side-dressed application after plants are established.

Side-dressed fertilizer. Usually, an incorporated fertilizer application typically banded to the side of crop rows after plant emergence. [also see *top-dressed*]

Slow-release (coated) fertilizer. A fertilizer term used interchangeably with delayed release, controlled release, controlled availability, slow acting, and metered release to designate a rate of dissolution much less than is obtained for completely water-soluble compounds. Slow release may involve either compounds that dissolve slowly or soluble compounds coated with substances relatively impermeable to water.

Slurry fertilizer. A fluid containing dissolved and undissolved plant nutrients. The suspension of the undissolved plant nutrients may be inherent with the materials or produced with the aid of a suspending agent of non-fertilizer properties. Mechanical agitation may be necessary in some cases to facilitate uniform suspension of undissolved plant nutrients.

Solution, fertilizer. A clear liquid fertilizer in which all nutrients have been completely dissolved in water.

Specialty fertilizer. A fertilizer distributed for non-farm use. Refers primarily to lawn and garden products.

Split-application. The fertilizer is added in two or more portions at different times during the growing season.

Starter (pop-up) fertilizer. A fertilizer applied in relatively small amounts with or near the seed, usually during planting for the purpose of accelerating early growth of the crop plants. Liquid or solid fertilizer placed near or in contact with the seed or the roots or new transplants, is commonly considered a starter fertilizer. Starter fertilizers provide high concentrations of nutrients near developing seedlings that can overcome nutrient uptake problems associated with low soil nutrient content, low soil temperature (e.g. with phosphorus), and compaction.

Suspension, fertilizer. A fluid containing dissolved and undissolved plant nutrient compounds. Suspension of the undissolved materials is usually produced with the aid of a suspending agent of non-fertilizer properties (e.g., clay). Mechanical or air agitation may be necessary to facilitate uniform suspension of undissolved plant nutrients.

Synthetic fertilizer. Any substance generated from another material or materials by means of a chemical reaction.

Top-dressed fertilizer. A non-incorporated surface application of fertilizer to a soil after the crop has been established. [also see *side-dressed*]

Ferrous sulfate ($FeSO_4 \cdot 7H_2O$). See *iron sulfate*.

Fibric soil material. In an unrubbed condition, organic fibers compose of 2/3 of the mass. [also see *hemic;* and *sapric*]

Fibrous root. A root system found in monocots in which branches develop from the adventitious roots, forming a system in which all roots are about the same size and length.

Field capacity (field water capacity; field moisture capacity). The content of water, on a mass or volume basis, remaining in a soil 2 or 3 days after having been wetted with water and after free drainage is negligible. (2) The percentage of water a particular soil will hold against the action of gravity. About half is available for crop usage. Estimated at –33 kPa water potential.

Soil will hold water against the pull of gravity keeping it available for plants to extract through their root zones. There are limits to the amount of available water. The upper limit is the field capacity (FC) of the soil. The lower limit is the permanent wilting point (PWP). [also see *permanent wilting point*]

Filler. Any substance added to fertilizer materials to provide bulk, prevent caking or serve some purpose other than providing essential plant nutrients.

Fish emulsion. Fertilizing material from which the guaranteed nutrients are derived primarily from fish, which contains a minimum of 40% total solids from fish, and which may contain additional sources of nitrogen, available phosphoric acid, and soluble potash for standardization purposes or stabilization purposes, or for both purposes, that shall be included in the required guaranteed analysis and derivation statement. [also see *tankage, fish*]

Fixation. The processes by which chemical elements are converted form a soluble or exchangeable form to a much less soluble or to a non-exchangeable form. Generally refers to reactions of phosphorus, ammonium, and potassium leading to decreased availability.

> **Ammonium.** Ammonium (NH_4^+) can be fixed by the same clay minerals that fix potassium, the in the same manner. [also see *reversion (of plant nutrients)*]
>
> **Phosphorus.** The more acid the soil and the higher its clay content, the greater its capacity for fixing phosphorus. On such soils, maintaining soil pH values at optimum (pH = 6.4) lessens fixation and improves nutrient use efficiency.
>
> **Potassium.** Fixed potassium ions are trapped between the silica sheets of certain soil clay minerals, especially vermiculite and illite. These trapped ions may be slowly released over time.

Fixation [of dinitrogen gas (N_2)]. See *dinitrogen fixation (also biological nitrogen fixation BNF)*.

Flocculant. Any of various soluble materials that have a combining effect of destroying the dispersed state of solids in solid-liquid mixtures. With clay dispersions in soils, almost any salt solution has the effect of a flocculant.

Flocculate (flocculation). The aggregation, or clumping together, of colloidal soil particles due to the ions in solution. In most soils the clays and humic substances remain flocculated due to the presence of doubly and triply charged *cations*. Higher levels of sodium ions (Na^+) causes soils to deflocculate. This problem can be corrected with the addition of calcium ions (Ca^{2+}) [doubly charged cations], especially from anhydrite or gypsum. [also see *deflocculate;* and *dispersion*]

Flood, 100-year. A 100-year flood does not refer to a flood that occurs once every 100 years, but to a flood level with a 1% chance of being equaled or exceeded in any given year.

Flood plain. The nearly level plain that borders a stream and is subject to inundation under flood-stage conditions unless protected artificially. It is usually a constructional landform built of sediment deposited during overflow and lateral migration of the streams.

Flora. (1) Term collectively applied to all of the plants in an area. The botanical counterpart of fauna. (2) A plant or plant life of a specific region or particular period. [also see *fauna*]

Fluvent. Floodplain soils; characterized by buried horizons and irregularly decreasing amounts of organic matter with depth.

Foliar diagnosis (plant tissue and/or petiole analyses). An estimate of a plant's nutrient deficiencies or sufficiencies by analytical measurement of selected parts of the plants, and the color and growth characteristics of the foliage of the plants.

Foliar spray. The application of liquid fertilizer to the foliage of plants, just enough to wet the leaves.

Footslope. The hillslope position that forms the inner, gently inclined surface at the base of a slope. In profile, footslopes are commonly concave and are situated between the *backslope* and the *toeslope*. [see *catina*; *backslope*; *shoulder*; and *toeslope*]

Forage. Plant materials that are used as feed for domestic animals. Forages may be grazed or cut for hay or silage.

Formation (or forming) factors, soil. The variables, usually interrelated natural agencies, that are active in and responsible for the formation of soil. The factors are usually grouped into five major categories as follows: parent material, climate, organisms, topography, and time.

Formula. See *chemical formula*.

Formula weight. The sum of the atomic weights of all atoms in a formula unit of a compound.

Example: The formula weight of anhydrite ($CaSO_4$)

Ca	=	40.078
S	=	32.066
O	=	63.997 (15.9994 times 4 = 63.997)
		136.142

Fossil fuels. Fossil fuels include coal, crude oils, oil shales, tar sands and natural gases such as butane, ethane, methane which occur naturally from the decomposition of plant and sea and land organisms over millions of years. These natural resources contain stored energy from the sun that is released upon combustion.

Fragipan. Brittle subsurface restricting soil horizon, usually loamy textured and weakly cemented.

Fragment, soil. A small mass of soil produced by a disturbance.

Friable. (1) A soil consistency term pertaining to the ease of crumbling of soils. (2) Generally refers to soils that are high in organic matter, often loamy, and easily worked with fingers. (3) A term applied to soils that when either wet or dry, crumbles easily between fingers. [also see *consistency, soil*]

Freely drained soil. A soil that allows water to percolate freely.

Freshwater. Water that contains less than 1,000 milligrams per liter (mg L^{-1}) [or parts per million] of dissolved solids. Generally, more than 500 mg L^{-1} of dissolved solids is undesirable for drinking and many industrial uses.

Frugivore. An animal which primarily eats fruit. Many bats and birds are frugivores.

Fruiting body. Macroscopic reproductive structure produced by some fungi, such as mushrooms, and some bacteria, including myxobacteria. Fruiting bodies are distinctive in size, shape, and coloration for each species.

Fulvic acid ($C_{14}H_{12}O_8$). The pigmented yellow organic material that remains in solution after removal of humic acid by acidification. An indefinite term for the mixture of organic substances remaining soluble after a soil extract, using dilute alkali, has been acidified. [also see *humic acid;* and *organic matter, soil*]

Fulvic acid fraction. The fraction of soil organic matter that is soluble in both alkali and dilute acid. [also see *organic matter, soil*]

Fungi (singular fungus). (1) Nonphototrophic, eukaryotic microorganisms that contain rigid cell walls. (2) Simple plants that lack a photosynthetic pigment. The individual cells have a nucleus surrounded by a membrane, and they may be linked together in long filaments called hyphae, which may grow together to form a visible body. The fungi, together with heterotrophic bacteria and a few other groups of heterotrophic organisms, are the <u>decomposers</u> of the biosphere. Their activities are as necessary to the continued existence of the world as are those of the food producers. [also see *bacteria, heterotrophic; and mycorrhiza*]

Saprophytic fungi. Fungi that decompose dead organic matter. Fungi tend to decompose more complex compounds such as fibrous plant residues and wood, while bacteria tend to decompose simpler organic compounds, such as root exudates and fresh plant residue.

Mycorrhizal fungi. Fungi that form symbiotic associations with plant roots. These fungi get energy from the plant (photosynthates) and help supply nutrients to the plant.

Fusiform. Spindle-shaped; a fusiform cell or other plant structure having a spindle-like shape that is wide in the middle and tapers at both ends. [Example: some roots]

Gaia hypothesis. (1) A hypothesis proposed during the early 1970s which states that all living organisms have the ability to affect their surroundings such as the atmosphere, lithosphere, and climate to maximize its biological success. The hypothesis connects the evolution and survival of a species to the evolution and conditions of its environment. (2) A hypothetical "superorganism" composed of the Earth's four spheres: the biosphere, hydrosphere, lithosphere, and atmosphere.

Gangue. In mining, gangue is the commercially worthless material that surrounds, or is closely mixed with, a wanted mineral in an ore deposit. The separation of mineral from gangue is known as mineral processing, mineral dressing or ore dressing and it is a necessary and often significant aspect of mining.

For any particular ore deposit, and at any particular point in time, the concentration of the wanted mineral(s) in the gangue material will determine whether it is commercially viable to mine that deposit.

Gastropod. A member of the Gastropoda class of mollusks that includes snails and slugs.

Gene. Unit of heredity; a segment of DNA specifying a particular protein or polypeptide chain, a tRNA or an mRNA.

Genus (plural genera). The first name of the scientific name (binomial); the taxon between family and species.

Geomorphology. The study of the origin of physical features of the Earth as they are related to geological structure and denudation.

Geosmin. Geosmin, which literally translates to "earth smell," is an organic compound produced by several classes of soil microbes, especially Streptomycetes (an Actinomycete). Geosmin is responsible for the "earthy" smell of freshly cultivated, healthy soil and for the earthy taste of beets. The human nose is extremely sensitive to geosmin and is able to detect it at concentrations as low as 5 parts per trillion. [also see *Actinomycete*]

Gley (gleying). A layer of mineral soil developed under conditions of poor drainage or otherwise poor (anaerobic) aeration; resulting in the reduction of certain elements including oxidized ferric iron [Fe^{3+}] to ferrous iron [Fe^{2+}]), and in gray and/or blue colors and mottles (blobs of variously colored soils). Rice fields in California typically have gleyed soil conditions after a period of flooding.

Glucose. A six-carbon single, simple sugar ($C_6H_{12}O_6$); the most common energy source and the primary product of photosynthesis. Glucose is polymerized (a dehydration reaction where water is removed) to make cellulose and chitin. Plants store glucose as the polysaccharide starch.

Granite. An igneous rock that contains quartz, feldspar, and varying amounts of biotite and muscovite.

Granular (soil structure). Soil structure in which the individual grains are grouped into spherical aggregates with indistinct sides. Highly porous granules are commonly called "crumbs." A well-granulated soil has the best structure for most common crop plant growth and production.

Granule. A natural soil ped or aggregate of relatively low porosity. [also see *aggregate*]

Grass cycling. The natural recycling of grass by leaving clippings on the lawn when mowing. Grass clippings will quickly decompose, returning valuable nutrients and organic matter back to the soil.

Grassland. A region in which the climate is dry for long periods of the summer, and freezes in the winter. Grasslands are characterized by grasses and other erect herbs, usually without trees or shrubs. Grasslands occur in the dry temperate interiors of continents, and first appeared in the Miocene period.

Gravel. Rounded or angular fragments of rock 2 millimeters to 7.6 millimeters (up to 3 inches) in diameter. An individual piece of gravel is a pebble.

Gravitational water. *See water, gravitational.*

Grazers. Organisms, such as protozoa, nematodes, and microarthropods, that feed on bacteria and fungi.

Green algae. See *algae, green.*

"Greenhouse" effect. Entrapment of solar radiant energy (heat), thus warming the Earth, by an increase in amounts of gases (carbon dioxide, methane, etc.) and water vapor in the atmosphere.

"Green-manure" crop. Any crop grown for the purpose of being turned under while green, or soon after maturity, for soil improvement, organic matter, and for the benefit of succeeding crops.

Green waste. A term used to refer to urban landscape waste generally consisting of leaves, grass clippings, weeds, yard trimmings, wood waste, branches/stumps, home garden residues, and other miscellaneous organic materials.

Grey water. Wastewater from clothes washing machines, showers, bathtubs, hand washing, lavatories and sinks.

Groundwater. (1) Water within the saturated zone of the Earth that supplies wells and springs and is free to move under the influence of gravity. (2) Water found underground as a result of rainfall, ice and snowmelt, submerged rivers, lakes, and springs. Groundwater carries minerals which can accumulate in the remains of buried organisms and eventually cause fossilization.

Groundwater, confined. Groundwater under pressure significantly greater than atmospheric, with its upper limit the bottom of a bed with hydraulic conductivity distinctly lower than that of the material in which the confined water occurs.

Groundwater, unconfined. Water in an aquifer that has a water table that is exposed to the atmosphere.

Groundwater recharge. Inflow of water to a groundwater reservoir from the surface. Infiltration of precipitation and its movement to the water table is one form of natural recharge. Also, the volume of water added by this process.

Groundwater table. The upper limit of the groundwater.

Growing season. The portion of the year when soil temperatures are above "biologic zero," 41 °F (4 °C), as defined by "Soil Taxonomy."

Guano. The decomposed (or partly decomposed) dried excrement of birds and bats used for fertilizer. Guano consists of approximately 10% nitrogen, 3% phosphorus, and 1% potassium, plus traces of other essential plant nutrients.

Guard cells. Specialized epidermal cells that flank stomata and whose opening and closing regulates gas exchange and water loss. [also see *epidermis (epidermal cells)* and *stomate*]

Guaranteed analysis (fertilizer).

Example of Guaranteed Fertilizer Analysis

SUPER SUPREME PLANT FOOD	
10-20-10	
Guaranteed Analysis	
Total Nitrogen (N)	10%
Available Phosphate (P_2O_5)	20%
Soluble Potash (K_2O)	10%
Net Wt. 50 lbs.	
Manufactured by: The Super Supreme Fertilizer Company Bumfucky, USA	

The guaranteed fertilizer analysis tells the user the guaranteed percentage of the nitrogen, phosphate and potash within the product. The product cannot contain more or less of a listed guarantee. To ensure the guaranteed analysis is correct, Trade and Consumer Protection randomly samples and tests fertilizers each year.

The analysis is usually designated as total nitrogen (N), available phosphate (P_2O_5), and soluble potash (K_2O). [Example: a 10-20-10 fertilizer grade contains 10% N, 20% P_2O_5, and 10% K_2O.

Gully. A small channel with steep sides caused by erosion and cut in unconsolidated materials by concentrated but intermittent flow of water usually during and immediately following heavy rains or ice/snow melt. A gully generally is an obstacle to wheeled vehicles and too deep (e.g., greater than 0.5 meters) to be obliterated by ordinary tillage. A rill is of lesser depth and can be "smoothed over" by ordinary tillage).

Gypsum (calcium sulfate dihydrate) ($CaSO_4 \cdot 2H_2O$). Gypsum, from the Greek "gypos" meaning plaster, is hydrated calcium sulfate and contains at least 70% $CaSO_4 \cdot 2H_2O$ (AAPFCO). Also called "land plaster" by Benjamin Franklin. Gypsum is a naturally occurring sedimentary evaporate mineral with a solubility of 0.205 grams per 100 grams water (2010 CRC Handbook of Chemistry and Physics). A typical analysis of high quality gypsum contains 22.5% calcium and 18.0% sulfur. Gypsum forms through direct precipitation from saline waters or through alteration of anhydrite. Same uses in agriculture and horticulture as anhydrite ($CaSO_4$). Often called the "universal soil amendment" since gypsum is used as a source of calcium and sulfur, and is widely used in reclaiming sodic soils, and correcting water infiltration problems associated with high levels of sodium and magnesium in soils and irrigation water. The addition of solution-grade calcium sulfate to irrigation water will also help replace any calcium precipitated as lime due to high concentrations of bicarbonate. [also see *anhydrite; hemihydrate; hydrate;* and *bicarbonate (HCO_3^-)*]

Gypsum requirement. The quantity of gypsum, or its equivalent (e.g., anhydrite), required reducing the exchangeable sodium fraction of a given amount of soil to an acceptable level where dispersion of soil colloids, phytotoxicity, and permeability problems do not take place. The gypsum requirement is expressed in tons of 100% gypsum per acre-foot of soil. Gypsum treatments may also be needed when soil electrical conductivity values are less than 0.60 decisiemens per meter.

Haber-Bosch process. An industrial process for the preparation of ammonia (NH_3) from nitrogen and hydrogen with a specially prepared iron-based catalyst, high temperature, and pressure.

$$3H_2 + N_2 \xrightarrow{\text{iron-based catalyst (1,200}^\circ \text{C, 500 atm.)}} 2NH_3 \text{ (ammonia)}$$

Habitat. The place and conditions in which an organism lives.

Halomorphic soil. A soil containing a significant proportion of soluble salts.

Halophile. (1) An organism that lives in areas of high salt concentration. (2) An organism requiring or tolerating a saline environment. (3) A group of archaebacteria that is able to tolerate high salt concentrations. These organisms must have special adaptations to permit them to survive under these conditions.

Halophyte. A plant that requires or tolerates a saline (high salt) environment.

Hardpan. A hardened or cemented soil horizon or layer. A soil layer with physical characteristics that limit root penetration and restrict water movement. [also see *pan*; *cemented (soil)*; *duripan*; *claypan*; and *horizon, soil*]

Hardness (water). A water-quality indication of the concentration of alkaline salts in water, mainly calcium and magnesium. If the water you use is "hard" then more soap, detergent, or shampoo is necessary to raise a lather.

Headwater. (1) The source and upper reaches of a stream; also the upper reaches of a reservoir. (2) The water upstream from a structure or point on a stream. (3) The small streams that come together to form a river. Also may be thought of as any and all parts of a river basin except the mainstream river and main tributaries.

Heavy metals. Naturally occurring metals in rocks, minerals, and soils having densities greater than 5.0 Mg per cubic meter (312 pounds per cubic foot). In soils, these include the elements cadmium (Cd), cobalt (Co), chromium (Cr), copper (Cu), iron (Fe), mercury (Hg), manganese (Mn), molybdenum (Mo), nickel (Ni), lead (Pb), and zinc (Zn). Many are toxic to plants and animals when accumulated in high, or in some cases, even low concentrations. [also see metal densities below]

Densities of Some Common Metals			
Metals	Specific gravity	Lbs in^{-3}	Lbs ft^{-3}
Aluminum	2.702	0.098	169
Brass	8.4 – 8.7	0.303 – 0.314	524 - 556
Bronze	7.4 - 8.9	0.267 – 0.322	462 - 556
Cobalt	8.9	0.322	556
Copper	8.93	0.323	557
Gold, pure	19.32	0.698	1,206
Gold, alloys	15.3 - 19.3	.553 - 0.698	955 - 1205
Iron, pure	7.86	0.284	491
Iron, wrought	7.4 - 7.9	0.275 - 0.285	474 - 493
Iron, cast	7.03 - 7.13	0.254 - 0.258	439 - 445
Lead	11.34	0.41	710
Magnesium	1.738	0.063	108
Manganese	7.35	0.263	455
Mercury	13.546	0.489	846
Nickel	8.9	0.322	556
Platinum	21.45	0.775	1,339
Plutonium	19.8	0.715	1236
Silver, pure	10.5	0.379	655
Silver, alloys	10 - 12	0.362 - 0.434	625 - 750
Steel	7.7 - 7.93	0.278 - 0.286	481 - 495
Tin	7.3	0.264	456
Titanium	4.5	0.163	281
Tungsten	19.3	0.697	1,205
Uranium	18.9	0.683	1180
Zinc	7.14	0.258	446

Compiled from Various Sources

Heavy soil (obsolete). A soil with high clay content, and which is difficult to cultivate.

Hemic soil material. In an unrubbed condition, 1/3 to 2/3 of the total mass is composed of organic fibers (intermediate in decomposition between fibric and sapric). [also see *fibric;* and *sapric*]

Hemihydrate (calcium sulfate hemihydrate) ($CaSO_4 \cdot nH_2O$). Hemihydrate is better known as plaster of Paris, while the dihydrate ($CaSO_4 \cdot 2H_2O$) occurs naturally as gypsum. The anhydrous form occurs naturally as anhydrite ($CaSO_4$). Depending on the method of calcination of calcium sulfate dihydrate, specific hemihydrates are sometimes distinguished: alpha-hemihydrate and beta-hemihydrate. The two forms appear to differ only in crystal size. Alpha-hemihydrate crystals are more prismatic than beta-hemihydrate crystals and when mixed with water form a much stronger and harder superstructure.

Heating gypsum to between 100 °C and 150 °C (302 °F) partially dehydrates the mineral by driving off approximately 75% of the water contained in its chemical structure. The temperature and time needed depend on ambient partial pressure of H_2O. Temperatures as high as 170 °C are used in industrial calcination, but at these temperatures γ-anhydrite begins to form. The reaction for the partial dehydration is:

$$CaSO_4 \cdot 2H_2O + heat \longrightarrow CaSO_4 \cdot \sim\tfrac{1}{2}H_2O + \sim 1\tfrac{1}{2}H_2O\ (steam)$$

The partially dehydrated mineral is called calcium sulfate hemihydrate or calcined gypsum (commonly known as plaster of Paris) ($CaSO_4 \cdot nH_2O$), where n is in the range 0.5 to 0.8. [also see *gypsum; anhydrite;* and *calcination (calcining)*]

Herb. (1) Generally any plant which does not produce wood and is not as large as a tree or shrub. (2) A non-woody seed plant with a relatively short-lived aerial portion.

Herbaceous. (1) An adjective referring to non-woody plants. (2) Term applied to a non-woody stem/plant with minimal secondary growth.

Herbaceous peat. An accumulation of organic material, decomposed to some degree, that is predominantly the remains of sedges, reeds, cattails, and other herbaceous plants.

Herbivore. Literally, an organism that eats plants or other autotrophic organisms. The term is used primarily to describe animals.

Heterotrophic bacteria. *See bacteria, heterotrophic.*

Heterotrophic nitrification. Biochemical oxidation of ammonium (NH_4^+) to nitrite (NO_2^-) and nitrate (NO_3^-) by heterotrophic microorganisms. [also see *nitrification*]

Heterotrophic organism. An organism capable of deriving energy for life processes only from the decomposition of organic compounds and incapable of using inorganic compounds as sole sources of energy or for organic synthesis. Heterotrophs, such as bacteria and fungi, cannot manufacture organic compounds and so must feed on organic materials that have originated in other plants and animals. Includes animals and fungi. [also see *autotrophic organism; fungi;* and *bacteria, heterotrophic*]

Histosol. A soil order in the taxonomic system that is composed of mucks and peats. Histosols have high concentrations of organic materials in the surface of the soil.

Holistic agriculture. South African statesman Jan Smuts first coined the term holistic in 1926. Holistic comes from the Greek word *holos* meaning all and everything. In his book "Holism and Evolution" Smuts made the following observations about how ecosystems function: (1) Nature only functions in whole within wholes; (2) Nature has no parts; (3) The whole is greater than the sum; (4) Nature will never be understood by studying its parts.

From a practical perspective, holistic farming is: (1) farming while still maintaining environmental and ethical integrity; (2) holism embraces the idea that everything is influenced by everything else. No person, family, farm, business, community or nation can operate in isolation from any other.

Holocene period. The Holocene epoch is the geological period extending from the present day back to about 10,000 radiocarbon years; approximately 11,430 (± 130 calendar years) before present.

Horizon, soil. A soil horizon is a specific layer in the soil which measures parallel to the soil surface and possesses physical characteristics which differ from the layers above and beneath.

Horizon formation is a function of a range of geological, chemical, and biological processes and occurs over long time periods. Soils vary in the degree to which horizons are expressed. Relatively new deposits of soil parent material, such as alluvium, sand dunes, or volcanic ash, may have no horizon formation, or only the distinct layers of deposition. As age increases, horizons generally are more easily observed. The exception occurs in some older soils, with few horizons expressed in deeply weathered soils, such as the oxisols in tropical areas with high annual precipitation.

Horticulture. The art and science of growing fruits, vegetables, and ornamental plants.

Host. An organism capable of supporting the growth of a virus or other parasite.

Hue. The dominant spectral color, and one of the three color variables. Hue is used to determine the color of soils. [also see *value;* and *chroma*]

Humates. Humates are the salts of humic acids collectively or the salts of humic acids specifically. [also see *humic acid; fulvic acid; ulmic acid; humin; humus;* and *organic matter, soil*]

Humic acid ($C_9H_9NO_6$). A mixture of dark-colored organic materials of indefinite composition that can be extracted from soil with dilute alkali and other reagents and that is precipitated by acidification to pH 1 to 2. [also see *humates; fulvic acid; ulmic acid; humin; humus;* and *organic matter, soil*]

Humic substances. (1) A series of relatively high-molecular-weight, brown-to-black substances formed by secondary synthesis reactions. (2) A general category of naturally occurring heterogeneous organic substances. The term is generic in a sense that it describes the colored material or its fractions obtained on the basis of solubility characteristics, such as humic acid or fulvic acid. [also see *humates; fulvic acid; ulmic acid; humin; humus;* and *organic matter, soil*]

Humidity. The moisture content of air. Relative humidity is the ratio of the amount of water vapor actually present in the air to the greatest amount possible at that same temperature.

Humification. (1) The process whereby the carbon of organic residues is transformed and converted to humic substances through biochemical and chemical processes. (2) The processes involved in the decomposition of organic matter and leading to the formation of humus.

Humilluvic material. Illuvial humus that accumulates after prolonged cultivation of some acid organic soil. [also see *acid sulfate soils;* and *illuviation*]

Humin. A dark-colored, bitter organic compound derived from humus. There is no chemical formula for humin since it is a complex organic compound, but is comprised of approximately: 45-55% carbon; 35-45% oxygen; 3-6% hydrogen; 1-5% nitrogen and 0-1% sulfur. [also see *fulvic acid; humic acid; ulmic acid; humus;* and *organic matter, soil*]

Humus. (1) The total of the organic compounds in soil exclusive of undecayed plant and animal tissues, their "partial decomposition" products, and the soil biomass. The term is often used synonymously with soil organic matter. (2) The major resistant fraction of organic matter in the soil; as opposed to detritus, the active fraction of soil organic matter. (3) The well-decomposed fraction of the soil organic matter remaining, usually amorphous and black colored, after the major portion of added residues and/or detritus has decayed.

Humus provides nutrients for plants and increases the water retention of soil. The importance of humus to the growth of plants is due principally to its high buffer capacity over a considerable range of pH values. Humus also tends to stabilize soil structure and has a high cation exchange capacity.

Although scientists have been intently studying humus for a century or more, they still do not know its chemical formula. It is certain that humus does not have a single chemical structure, but is a very complex mixture of similar substances that vary according to the types of organic matter that decayed and the environmental conditions and specific organisms that made the humus. [also see *buffer capacity; detritus; fulvic acid; humin; humic acid; organic matter, soil;* and *cation exchange capacity*]

Hydrate (noun). A term used in inorganic chemistry to indicate a compound formed by the chemical combination of water and some other substance. Gypsum ($CaSO_4 \cdot 2H_2O$) is a hydrate of water and the mineral anhydrite ($CaSO_4$).

Hydrate (verb). To cause to take up or combine with water or the elements of water, especially to form a hydrate (noun).

Hydrated lime [$Ca(OH)_2$]. A strong base similar to household lye. Hydrated lime, obtained when calcium oxide is missed (or slaked) with water, is a dry product primarily consisting of calcium (and sometimes magnesium) hydroxides; and is used as an agricultural liming product with a 120 to 136% calcium carbonate equivalent.
When heated to 512 °C calcium hydroxide decomposes into calcium oxide (CaO) and water.

$$Ca(OH)_2 \xrightarrow{heat\ (512\ °C)} CaO + H_2O$$

[also see *lime, agricultural*]

Hydration. The process whereby a substance takes up water. Chemical combination with water. [also see *hydrate (noun)*; and *hydrate (verb)*]

Hydraulic conductivity. An expression of the readiness with which a liquid such as water flows through a solid such as soil in response to a give potential gradient.

Hydric layer. This is a layer of water that extends from a depth of not less than 40 centimeters (approximately 101 inches) from the organic surface to a depth of more than 1.6 meters (approximately 1.75 yards).

Hydric soils. Soils that are saturated with water long enough during the growing season to become anaerobic. The soils will usually be characterized by anaerobic soil zones and wetland vegetation.

Hydrocarbons. Chemicals containing only carbon and hydrogen. These are of prime economic importance because they encompass the constituents of the major fossil fuels, petroleum and natural gas, as well as plastics, waxes, and oils. The simplest hydrocarbon is methane (CH_4).

Hydrogen (H). Hydrogen, an essential plant nutrient, is absorbed by plant roots as water (H_2O). In addition to carbon, nearly all organic compounds contain hydrogen including carbohydrates, proteins, nucleic acids and lipids. For example, hydrogen comprises approximately 8.7% of the fresh weight of alfalfa plants. [also see *plant nutrients, essential*]

Hydrogen bond. Chemical bond between a hydrogen atom of one molecule and two unshared electrons of another molecule.

Hydrologic cycle. The recycling of water on Earth. There are five steps: (1) Condensation occurs when water vapor in the atmosphere turns into a liquid and forms clouds. (2) The clouds release precipitation in the form of rain, snow, or sleet when the moisture in the clouds becomes too heavy. (3) Infiltration occurs when the water seeps into the ground. The infiltration rate depends on the permeability of the ground. (4) When there is too much water on the surface to infiltrate into the ground, it becomes runoff and ends up in lakes, streams, or oceans. (5) Evapotranspiration is a two-part process driven by the sun. Evaporation occurs when water is turned into a vapor from the ground and transpiration is water loss by plants. The water vapor rises and starts the process over again. [also see *water cycle*]

Hydrolysis. Hydro (water) and -lysis (splitting). (1) The reaction of an ion (splitting of one molecule into two) with the addition of water to produce soluble salts and either hydronium (H_3O^+) [hydrogen, H^+] ion or hydroxide (OH^-) ion. (2) The process by which fertilizer salts react with water.

Hydromorphic soils. Soils developed in the presence of excess water.

Hydronium ion. The H_3O^+ ion; also called the hydrogen ion and written H^+ (aqueous).

Hydrophilic. Water loving. (1) Molecules that readily dissolve in water are called hydrophilic. Such molecules slip into aqueous solution easily because their partially charged regions attract water molecules and thus compete with the attraction between the water molecules themselves. (2) A term applied to polar molecules that can form a hydrogen bond with water. [also see *hydrophobic (water fearing)*]

Hydrophobic. Water fearing. (1) A substance having a strong tendency to reject association with water. (2) Hydrophobic compounds do not dissolve easily in water and are usually non-polar. Oils and other long hydrocarbons are hydrophobic. Molecules that lack polar regions, such as fats, also tend to be insoluble in water. As a result, non-polar molecules tend to cluster together in water, just as droplets of fats tend to coalesce on the surface of, e.g., chicken soup. [also see *hydrophilic (water loving)*]

Hydrophobicity, soil (hydrophobic soils). Soils that repel water are considered hydrophobic, the tendency for a soil particle or soil mass to resist hydration. A thin layer of soil at or below the mineral soil surface can become hydrophobic after intense heating. The hydrophobic layer is the result of waxy substances that generally are derived from plant material burned during a hot fire. The waxy substance penetrates into the soil as a gas and solidifies after it cools, forming a waxy coating around soil particles. Hydrophobic soils repel water reducing the amount of water infiltration. Decreased infiltration into the soil results in increased erosion with greater amounts of runoff, where much of the fertile topsoil layer can be lost.

Hydrophytic (plants). (1) Aquatic plants that have adapted to living within aquatic environments that is at least periodically deficient in oxygen as a result of excessive water content. They are also referred to as hydrophytes or aquatic macrophytes. These plants require special adaptations for living submerged in water or at the water's surface. Aquatic plants can only grow in water or in soil that is permanently saturated with water. Aquatic vascular plants can be ferns or angiosperms (from a variety of families, including among the monocots and dicots). Seaweeds are not vascular plants but multicellular marine algae.

Agronomic plants that are hydrophytic include: wild rice (*Zizania* sp), Chinese water chestnut ((*Eleocharis dulcis*), and taro (*Colocasia esculenta*). Rice (*Oryza*) is originally not an aquatic plant.

Hydroponics. (1) The production of plants in a liquid solution or gravel medium supplemented with all required water-soluble nutrients for proper growth. (2) A system in which water-soluble nutrients are placed in intimate contact with the plant's root system, being grown in an inert supportive medium which supplies physical support for the roots but with does not add or subtract plant nutrients (AAPFCO).

Hydrosphere. That part of the atmosphere that contains water in the liquid, solid, or gaseous phase.

Hygroscopic. Having the capability of absorbing moisture from an atmosphere of high relative humidity. Soils and certain fertilizer salts and soil amendment materials have this potential.

Hygroscopic water. (1) Water that is adsorbed onto a surface from the atmosphere. (2) Water adsorbed by a dry soil from an atmosphere of high relative humidity.

Hygroscopicity. The tendency of salts to adsorb water whenever the vapor pressure of moisture in the air exceeds that of a saturated solution of the salt.

Hyperparasite. A parasite that feeds on another parasite.

Hyphae (singular hypha). (1) Any of the thread-like filaments of vegetative mycelium that constitute the body (mycelium) of a fungus. (2) Long and often branched tubular filament that constitutes the vegetative body of many fungi and fungus-like organisms. Bacteria of the order Actinomycetes also produce branched hyphae. [also see *mycelium;* and *mold*]

Igneous rock. Rock formed by solidification from a molten or partially molten material. [Examples: granite and basalt] [also see *sedimentary;* and *metamorphic*]

Illuvial soil horizon. A soil horizon that receives material in solution or suspension from some other part of the soil. [also see *eluvial soil horizon*]

Illuviation. The process of deposition of soil material removed from one horizon to another in the soil; usually from an upper to a lower horizon in the soil profile. [also see *eluviation*]

Imbibition. (1) In chemistry, the act of "imbibing." Absorption of fluid by a solid or colloid that results in swelling. (2) Adsorption of water and swelling of colloidal materials because of the adsorption of water molecules onto the internal surfaces of the materials.

Immobilization. The conversion of an element from the inorganic to the organic form in microbial or plant tissues. Example: If decomposing organic matter contains low N relative to C, the microorganisms will immobilize ammonium (NH_4^+) and nitrate (NO_3^-) in the soil. Significant amounts of soil NH_4^+ and NO_3^- are used by microorganisms (now incorporated as organic N) and becomes part of the soil's organic material. Microbes need N in a C:N ratio of about 8:1. [also see *C:N ratio*]

Impeded drainage. A condition that hinders the movement of water through soils under the influence of gravity. Impeded drainage can often be corrected/alleviated with the addition of soil amendments such as calcium sulfate (anhydrite and/or gypsum). [also see *anhydrite; and gypsum*]

Impermeable layer. A layer of solid material, such as rock or clay, which does not allow water to pass through.

Impervious. Resistant to penetration by fluids (e.g., water) or by roots.

Indigenous. Native to an area.

Indurated. A very strongly cemented soil horizon that will not soften on wetting. For example: an indurated duripan. [also see *duripan*]

Infiltration rate. A soil characteristic determining or describing the maximum rate at which water can enter the soil under specified condition, including the presence of an excess of water.

Infiltration, water. The entry of water downward into the soil surface. Factors that control the rate of water movement into the soil include: (a) percentage of calcium in relation to sodium plus magnesium of the basic cation saturation [there needs to be 8 times more calcium than magnesium plus sodium combined], (b) percentage of sand, silt, and clay, (c) soil structure, (d) organic matter, (e) depth of the soil to a pan layer, or other impervious layers, (f) amount of water in the soil, (g) soil temperature, and (h) soil compaction. [also see *percolation*; *leaching*; *basic cation saturation*; *anhydrite;* and *gypsum*]

Ingestion. The intake of water or food particles by "swallowing" them into the body cavity or into a vacuole. [also see *absorption*]

Inhibition. The prevention of growth or function.

Inoculate. To treat (usually seeds, but also applies to soil or culture media) with microorganisms to create a favorable response. [Example: legume seeds treated with *Rhizobia* to stimulate dinitrogen fixation. Also refers to the introduction of microbial cultures into growth medium]. [also see *dinitrogen fixation*]

Inoculum. Material used to introduce a microorganism into a suitable situation for growth.

Inorganic (compounds). (1) Inorganic compounds can be formally defined with reference to what they are not...organic compounds. Organic compounds contain carbon bonds in which at least one carbon atom is covalently linked to an atom of another type (commonly hydrogen, oxygen or nitrogen). Some carbon-containing compounds are traditionally considered inorganic. When considering inorganic chemistry and life, it is helpful to remember that many species in nature are not compounds per se, but are ions. (2) All chemical compounds in nature except compounds of carbon other than carbon monoxide, carbon dioxide, and carbonates. Substances in which carbon-to-carbon bonds are absent. [Examples: ammonium sulfate ((NH_4)$_2SO_4$), magnesium phosphate ($Mg_3(PO_4)_2$) and potassium nitrate (KNO_3)]. [also see *ion(s);* and *organic*]

Insoluble. Not soluble or capable of being dissolved. As applied to potassium and nitrogen, it refers to insolubility in water. As applied to phosphates in fertilizer, it refers to that portion of the total that is insoluble in both water and neutral ammonium citrate solution.

Intergrade. In soil science, a soil which contains the properties of two distinctive and genetically different soils.

Integrated Pest Management (IPM). (1) An ecosystem-based strategy that focuses on long-term prevention of pests or their damage through a combination of techniques such as biological control, habitat manipulation, modification of cultural practices, and use of resistant varieties. Pesticides are used only after monitoring indicates they are needed according to established guidelines, and treatments are made with the goal of removing only the target organisms. (2) The many uses of all available biological, physical, and chemical controls and crop rotation to reduce losses to crops caused by pests. [Examples include: uses of resistant plant varieties, natural predators of the pest, preventive measures, and judicious management practices such as crop rotations]

Interveinal chlorosis. Interveinal chlorosis is a yellowing of the leaves between the veins with the veins themselves remaining green. In grasses, this results in a striped effect. While there are several possible causes, this symptom frequently indicates a nutritional imbalance or deficiency. Manganese, iron, and zinc deficiencies all produce interveinal chlorosis. [also see *chlorosis;* and *necrosis*]

Intracellular. Existing, occurring, or functioning within a cell.

Intrusion. Magma (and the rock it forms) that has pushed into pre-existing rock. [also see *extrusive*]

Inversion (thermal). An anomaly in the normal positive atmospheric lapse rate (change of temperature with increasing altitude). This usually refers to a thermal inversion where temperature of the atmosphere increases rather than decreases with height.

In situ **(Latin).** In position. In its original place.

In vitro **(Latin).** Literally "in glass;" it describes whatever happens in a test tube or other receptacle, as opposed to *in vivo*. When a study or an experiment is done outside the living organism, in test tube, it is done *in vitro*.

In vivo **(Latin).** In the body, in a living organism, as opposed to *in vitro*. When a study or an experiment is done in the living organism, it is done *in vivo*.

Ionic bond. A type of chemical bond that involves a metal and a nonmetal ion (or polyatomic ions such as ammonium) through electrostatic attraction. A bond formed by the attraction between two oppositely charged ions. (1) A chemical bond formed by the electrostatic attraction between positive and negative ions. (2) A chemical bond in which atoms of opposite charge are held together by electrostatic attraction.

Example: $Na + Cl \longrightarrow Na^+ + Cl^- \longrightarrow NaCl$

Ion(s). An atom or a group of atoms whose negative or positive electric charge results from having lost or gained one or more electrons. When an acid, base, or salt dissolves in water, some of its atoms or elements separate into positive and negative ions. Cations are positive ions formed by the loss of electrons; anions are negative ions formed by the gain of electrons. The number of electrons lost or gained is denoted by a positive sign for cations (e.g., Mg^{2+} for magnesium) or a minus sign for an anion (e.g., Cl^-

for chloride). Most plant nutrients are absorbed by plant roots in the ionic form (NO_3^-, Ca^{2+}, Zn^{2+}, etc.). Anions and cations are <u>always</u> present in the soil solution. [also see *cations;* and *anions*]

Common Ions in Soil Science	
Positive ions (cations)	**Negative ions (anions)**
aluminum Al^{3+}	acetate $C_2H_3O_2^-$
ammonium NH_4^+	borate BO_3^{3-}
barium Ba^{2+}	bromide Br^-
calcium Ca^{2+}	carbonate CO_3^{2-}
chromium (II) Cr^{2+} chromous	hydrogen carbonate HCO_3^- (bicarbonate)
chromium (III) Cr^{3+} chromic	chlorate ClO_3^-
cobalt (II) Co^{2+} cobaltous	chloride Cl^-
cobalt (III) Co^{3+} cobaltic	chlorite ClO_3^-
copper (I) Cu^+ cuprous	chromate CrO_4^{2-}
copper (II) Cu^{2+} cupric	cyanide CN^-
gold Au^+	dichromate $Cr_2O_7^{2-}$
hydrogen H^+	dihydrogen phosphate $H_2PO_4^-$
iron (II) Fe^{2+} ferrous	fluoride F^-
iron (III) Fe^{3+} ferric	hydrogen oxalate $HC_2O_4^-$ (bioxalate)
lead (II) Pb^{2+} plumbous	hydrogen sulfate HSO_4^- (bisulfate)
lead (IV) Pb^{4+} plumbic	hydrogen sulfide HS^- (bisulfide)

lithium	Li$^+$	hydrogen sulfite	HSO$_3^-$ (bisulfite)
magnesium	Mg^{2+}	hydroxide	OH$^-$
manganese (II)	Mn^{2+} manganous	hydride	H$^-$
manganese (IV)	Mn^{4+} manganic	hypochlorite	ClO$^-$
mercury (I)	Hg$_2^{2+}$ mercurous	iodide	I$^-$
mercury (II)	Hg^{2+} mercuric	molybdate	MoO$_4^{2-}$
nickel (II)	Ni^{2+} nickelous	monohydrogen phosphate	HPO$_4^{2-}$
nickel (III)	Ni^{3+} nickelic	nitrate	NO$_3^-$
potassium	K$^+$	nitride	N^{3-}
silver	Ag$^+$	nitrite	NO$_2^-$
sodium	Na$^+$	oxalate	C$_2$O$_4^{2-}$
strontium	Sr^{2+}	oxide	O^{2-}
tin (II)	Sn^{2+} stannous	perchlorate	ClO$_4^{2-}$
tin (IV)	Sn^{4+} stannic	permanganate	MnO$_4^-$
zinc	Zn^{2+}	phosphate	PO$_4^{3-}$
		phosphite	PO$_3^{3-}$
		sulfate	SO$_4^{2-}$
		sulfide	S^{2-}
		sulfite	SO$_3^{2-}$

Compiled From Various Sources

Iron (Fe). An essential metallic micronutrient that is primarily absorbed by plants in the ionic form as the reduced ferrous ion (Fe^{2+}).

Most minerals soils consist of approximately 5.0% total iron, while agricultural crops require less than 0.5 parts per million in the soil. Therefore, any problem of iron supply to crops from soil is always one of availability. [also see *plant nutrient, essential;* and *cation*]

Iron is a catalyst in chlorophyll formation and acts as an oxygen carrier. It also helps form certain respiratory enzyme systems in the plant. Fe^{2+} is very immobile in plants and deficiencies show up first in the younger leaves as interveinal chlorosis (very light pale leaf color with veins remaining green). Factors contributing to iron deficiency include high availability of soil phosphorus, wet and cold soil conditions, high levels of other heavy metals, and low soil organic matter. The solution pH also has a dominant influence on iron solubility. While soil pH values of 6-7 are optimum for iron, iron solubility decreases about a thousand fold per pH unit rise above optimum. Conversely, at lower pH values and in anaerobic soils, iron can be present in toxic levels. Iron toxicity is particularly a problem in flooded rice soils since within a few weeks, flooding may increase the level of soluble Fe^{2+} from 0.1 ppm to 50-100 ppm. [also see *cations;* and *plant nutrients, essential*]

Iron sulfate (ferrous sulfate) ($FeSO_4 \cdot 7H_2O$). A greenish crystalline water-soluble compound with approximately 20% iron used as a fertilizer. Reduced ferrous sulfate has a solubility of 242 lbs/100 gallons of cold water.

Irrigation. (1) Intentional watering of the soil. (2) Irrigation is the controlled application of water for agricultural purposes through manmade systems to supply water requirements not satisfied by rainfall. Crop irrigation is vital throughout the world in order to provide the world's ever-growing populations with enough food. Many different irrigation methods are used worldwide, including:
Center Pivot: Automated sprinkler irrigation achieved by automatically rotating the sprinkler pipe or boom, supplying water to the sprinkler heads or nozzles, as a radius from the center of the field to be irrigated. Water is delivered to the center or pivot point of the system. The pipe is supported above the crop by towers at fixed spacings and propelled by pneumatic, mechanical, hydraulic, or electric power on wheels or skids in fixed circular paths at uniform angular speeds. Water is applied at a uniform rate by progressive increase of nozzle size from the pivot to the end of the line. The depth of water applied is determined by the rate of travel of the system. Single units are

ordinarily about 1,250 to 1,300 feet long and irrigate about a 130 acre circular area.

Drip: A planned irrigation system in which water is applied directly to the root zone of plants by means of applicators (orifices, emitters, porous tubing, perforated pipe, etc.) operated under low pressure with the applicators being placed either on or below the surface of the ground.

Flood: The application of irrigation water where the entire surface of the soil is covered by ponded water.

Furrow: A partial surface flooding method of irrigation normally used with clean-tilled crops where water is applied in furrows or rows of sufficient capacity to contain the designed irrigation system.

Gravity: Irrigation in which the water is not pumped but flows and is distribute by gravity.

Rotation: A system by which irrigators receive an allotted quantity of water, not a continuous rate, but at stated intervals.

Sprinkler: A planned irrigation system in which water is applied by means of perforated pipes or nozzles operated under pressure so as to form a spray pattern.

Subirrigation: Applying irrigation water below the ground surface either by raising the water table within or near the root zone, or by using a buried perforated or porous pipe system that discharges directly into the root zone.

Supplemental: Irrigation to ensure increased crop production in areas where rainfall normally supplies most of the moisture needed.

Surface: Irrigation where the soil surface is used as a conduit, as in furrow and border irrigation as opposed to sprinkler irrigation or subirrigation.

Traveling Gun: Sprinkler irrigation system consisting of a single large nozzle that rotates and is self-propelled. The name refers to the fact that the base is on wheels and can be moved by the irrigator or affixed to a guide wire.

Isomorphous substitution (replacement). The replacement of one atom by another of similar size in a crystal structure without disrupting or seriously changing the structure. When a substituting cation is of a smaller valence than the cation it is replacing, there is a net negative charge on the structure. The excess negative charge creates sites that attract cations from the surrounding solution and are held somewhat loosely. The negatively charged locations are called cation exchange sites. [also see *cation exchange*; *cation exchange capacity;* and *cation*]

Isotope. (1) One of several possible forms of a chemical element that differ from other forms in the number of neutrons in the atomic nucleus, but not in the number of protons, or chemical properties. (2) Two or more forms of an element that have the same atomic number but different numbers of neutrons in the atomic nucleus and therefore a different atomic weight. Most elements occur naturally as mixtures of isotopes. The symbol for an isotope usually consist of the mass number placed as a leading superscript before the elemental symbol. Long-lived isotopes such as carbon-14 (written: ^{14}C) are used to determine the age of objects that contained living matter.

Joule. The international system (SI) energy unit defined as a force of one Newton applied over a distance of one meter. 1 joule = 0.239 calorie.

K_2O. See *potash (K_2O, potassium oxide)*.

Kaolinite. A hydrous aluminosilicate clay mineral with the chemical composition $Al_2Si_2O_5(OH)_4$. It is a layered silicate mineral with one silicon tetrahedral sheet and one aluminum oxide-hydroxide octahedral sheet (a 1:1 crystal structure group). Rocks that are rich in kaolinite are known as china clay, white clay, or kaolin. The name is derived from Chinese *Gaoling or Kao-ling* ("High Hill") in Jiangxi province, China. In the Ione, California area, (Amador County) kaolinite is mined for use as pottery/china clay. [also see *montmorillonite*]

Kelp.

Typical Nutrients and Elements in Kelp Meal (*Ascophyllum nodosum*) (dry matter basis)			
Minerals		Trace elements	
Chlorine (Cl)	6.5%	Iodine (I)	927 ppm
Sodium (Na)	4.21%	Iron (Fe)	598 ppm
Sulfur (S)	2.8%	Aluminum (Al)	289 ppm
Potassium (K)	2.5%	Total arsenic (As)	23 ppm
Calcium (Ca)	1.7%	Manganese (Mn)	33 ppm
Nitrogen (N)	1.4%	Tin (Sn)	6.5 ppm
Magnesium (Mg)	0.76%	Selenium (Se)	4.9 ppm
Phosphorus (P_2O_5)	0.28%	Cobalt (Co)	4.5 ppm
Phosphorus (P)	0.15%	Copper (Cu)	4.0 ppm

Compiled From Various Sources

(1) A common name for any of the larger members of the order *Laminarialea* of the brown algae. (2) Any of several species of seaweed including Norwegian kelp (*Ascophyllum nodosum*); generally produced from saltwater farms and harvested for use as a certified organic fertilizer. Dried kelp will typically contain 1% to 4% nitrogen, 1% to 2% P_2O_5, and 1% to 5% K_2O.

Kelp is considered the richest marine plant available for agricultural use. For kelp meal, the seaweed is dehydrated immediately after harvesting to a specified moisture content by sun-drying and/or mechanical dehydration, then ground into meal to a specified particle size. Seaweed extract is produced from fresh, live plants which are processed into a soluble powder or liquid concentrated seaweed extract. Principally, kelp is a source of potassium, naturally chelated micronutrients, natural plant growth regulators, and organic matter. [also see *algae, brown;* and *seaweed*]

Kilogram. The international system (SI) base unit for mass; equal to 2.2046 pounds.

Klusterfuk. A situation of total disarray and confusion; usually associated with inept or brainless workers and/or personnel (e.g., government bureaucrats and college/university administrators). May be related or linked to panic, hysteria and madness.

Krotovina. See *crotovina*.

Labile (nutrient). (1) The total sum of a nutrient that readily solubilizes or exchanges to become available to plants during a season. (2) The term applied to denote an element that can be solubilized in a relatively short period of time. A labile nutrient may not directly available, but will be released or plant available relatively quickly.

Lacustrine. Pertaining to fresh water lakes.

Lacustrine deposit. Sediments deposited in fresh (non-saline) lake water and later exposed either by lowering of the water level or by the elevation of the land. [Example: the Tulare Lake basin of California's San Joaquin Valley]

Lamellae (singular lamella). (1) A thin layer, plate-like arrangement or membrane. (2) Layers of protoplasmic membranes within the chloroplast that contain the photosynthetic pigments.

Landform. A discernible natural landscape, such as a floodplain, stream terrace, or alluvial fan.

Land-use planning. The development of plans for the uses of land that, over long periods, will best serve the general welfare, together with the formulation of ways and means for achieving such uses.

Larva. Among invertebrates, an immature stage in the life cycle which usually is much smaller than, and morphologically different from, the adult. In insects with metamorphosis, the larva must become a pupa before reaching adulthood. [also see *metamorphosis;* and *pupa*]

Lateral root. A root that arises from another, older root. Also call a branch root, or secondary root, if the older root is the primary root.

Lattice structure. The orderly arrangement of atoms in crystalline material.

"LB" urea. See *low biuret urea.*

Leachate. Liquids that have percolated through a soil and that contain substances in solution or suspension.

Leaching. The downward removal of soluble or suspended materials by water movement or passage through the soil profile (percolation). In agriculture, leaching refers to the downward movement of free water (percolation) out of the plant root zone (rhizosphere). Leaching of nutrients, particularly nitrate (NO_3^-) leaching, can cause decreased nutrient use efficiency, lower yields, and environmental problems including nutrient accumulation in groundwater. [also see *infiltration, water;* and *percolation* (of *soil water)*]

Leaching requirement. The portion of the water entering the soil that must pass through the root zone in order to help prevent soil salinity from exceeding a specified value.

Leaflet. One of the parts of a compound leaf.

Leghemoglobin. Iron-containing, red pigment(s) produced in root nodules during the symbiotic association between *Rhizobia* and leguminous plants. The pigment is similar but not identical to mammalian hemoglobin.

Legume. Any member of the legume or pulse family *Leguminosae*. Legumes are characterized botanically by a fruit called a legume, or pod, that opens along two "sutures" when ripe. Legumes, in symbiotic relationship with *Rhizobia* bacteria, fix atmospheric nitrogen (N_2) in nodules on the plant's roots. It is estimated that from 100 to 300 pounds of nitrogen per acre may be fixed annually by a leguminous crop. [Examples: peas, beans, peanuts, clover, alfalfas, sweet clovers, soybeans, vetches, lespedezas, and kudzu]. [also see *nodule*; *Rhizobia;* and *symbiotic bacteria*]

Lentic waters. Ponds or lakes (standing water).

Leveled land. A land area, usually a field, that has been mechanically flattened or smoothed to facilitate management practices such as flood irrigation; as a result the natural soil has been partially or completely modified.

Lichen. A symbiotic relationship of a fungus and an alga whereby the fungus supplies water and dissolved nutrients and the alga photosynthesizes carbohydrates and fixes nitrogen. Lichens colonize bare minerals, rocks, and trees, a first step in rock weathering and soil formation since lichens exude carbonic acid (H_2CO_3) via respiration. The chemical mechanism of lichen weathering, or deteriorating $CaCO_3$ (limestone) through carbonic acid exuded from respiration, is shown below:

$$CaCO_3 + H_2CO_3 \text{ (carbonic acid)} \longrightarrow Ca(HCO_3)_2 \text{ (calcium bicarbonate)}$$

Liebig's law (of the minimum). A principle developed in agricultural science by Carl Sprengel (1828) and later popularized by Justus von Liebig. It states that growth is controlled not by the total of resources available, but by the scarcest resource (limiting factor). This concept was originally applied to plant or crop growth, where it was found that increasing the amount of plentiful nutrients did not increase plant growth. Only by increasing the amount of the limiting nutrient (the one most scarce in relation to "need") was the growth of a plant or crop improved. The growth and reproduction of an organism are determined by the nutrient substance (calcium, nitrogen, potassium, etc.) that is available in minimum quantity, the limiting factor.

Life cycle. The entire sequence of phases in the growth and development of any organism from time of zygote formation to gamete formation.

Ligand. An organic molecule that can bond to metals through two or more bonds. The ligand-metal is called a chelate. [also see *chelate*]

Light soil (obsolete). A soil that has a course texture and is easily cultivated. A sandy soil. [also see *heavy soil (obsolete)*]

Lignin ($C_9H_{10}O_2$, $C_{10}H_{12}O_3$, $C_{11}H_{14}O_4$). (1) A polymer in the secondary cell wall of woody plant cells that helps to strengthen and stiffen the wall. (2) One of the most important constituents of the secondary wall of vascular plants. (3) Cell wall material that helps cement cells together. After cellulose, lignin is the most abundant plant polymer. Lignin is very resistant to decomposition and constitutes much of the residual humus in soils. [also see *cellulose*; *humus*; and *organic matter, soil*].

Lime, agricultural. (1) In strict chemical terms, calcium oxide (CaO). (2) In practical agricultural/horticultural terms, a soil amendment containing calcium carbonate ($CaCO_3$), magnesium carbonate ($MgCO_3$), and oxides and/or hydroxides of calcium and/or magnesium, used to neutralize soil acidity and furnish calcium and/or magnesium for plant growth. Classification including calcium carbonate equivalent and limits in lime particle size is usually prescribed by law or regulation. [also see *lime requirement*; *limestone*; and *dolomitic lime*]

The chemical reactions for the lime neutralization process is:

$$H^+ - Clay + CaCO_3 \longrightarrow Ca^{++} - Clay + H_2CO_3 \text{ (carbonic acid)}$$

$$H_2CO_3 \longrightarrow H_2O + CO_2 \text{ (carbon dioxide)}$$

Lime requirement. The amount of agricultural liming material, as calcium carbonate equivalent, required changing a volume of soil to a specified state with respect to pH or soluble aluminum content. Generally, lime requirements are given in tons per acre of limestone ground finely enough so that all passes through a 10-mesh U.S. sieve [Note: 10-mesh U.S. sieve = 2,000 microns = 2.0 millimeters = 0.0787 inches]

Limestone ($CaCO_3$). A sedimentary rock composed of more than 50% of the mineral calcite, calcium carbonate ($CaCO_3$). If dolomite minerals [$CaMg(CO_3)_2$] are present in appreciable quantities, it is called a dolomitic limestone. [also see *lime, agricultural;* and *dolomitic lime*]

Liming Material (as $CaCO_3$ equivalent) Required to Raise pH of Top Seven Inches of Soil		
Soil Texture	Lime Requirements (tons per acre)	
	From pH 4.5 to 5.5	From pH 5.5 to 6.5
Sand and Loamy Sand	0.5	0.6
Sandy Loam	0.8	1.3
Loam	1.2	1.7
Silt Loam	1.5	2.0
Clay Loam	1.9	2.3
Organic Soil (muck)	3.8	4.3

USDA Handbook No. 18. 1993.

Common Liming Materials		
Name	Chemical Formula	$CaCO_3$ Equivalent (%)
Limestone	$CaCO_3$	100
Dolomitic lime	$CaCO_3 \cdot MgCO_3$	109 - 110
Burned or Quick Lime	CaO	150 - 179
Calcium silicate	$CaSiO_3$	60 - 90
Hydrated lime	$Ca(OH)_2$	120 - 136
Sugar beet lime	$CaCO_3$	80 - 90

Miller and Gardiner. Soils in Our Environment. 11th ed. 2007.

Effect of pH on solubility of CaCO₃ in Water	
pH	Solubility of CaCO$_3$ meq L^{-1}
6.21	19.3
6.50	14.4
7.12	7.10
7.85	2.70
8.60	1.10
9.20	0.80
10.12	0.40

Compiled From Various Sources

Limnic layer. This is a layer or layers, 5 centimeters (approximately 2 inches) or more thick, of coprogeneous earth (sedimentary peat), diatomaceous earth, or marl. Except for some of the coprogeneous earth containing more than 30% organic matter, most of these limnic materials are inorganic.

Limnic soil material. Organic or inorganic materials deposited in water by the action of aquatic organisms or derived from underwater and floating organisms. Marl, diatomaceous earth, and sedimentary peat (coprogeneous earth) are considered limnic materials.

Limnology. The study of river system ecology and life.

Lipid. (Greek *lipos* = fat). A generic term for all fats, oils, waxes, and related fatty compounds.

Liquid Fertilizer. See *fertilizer, liquid*.

Liquor ammonia. See *ammonia, liquor*.

LISA (obsolete). Acronym for "Low Input Sustainable Agriculture." The term has been replaced by "Sustainable Agriculture."

Lithic layer. This is a consolidated mineral layer (bedrock) occurring within 10-160 centimeters (approximately 4 to 63 inches) of the surface of organic soils.

Lithology. The branch of mineralogy that deals with the study of rocks and their composition.

Lithosphere. (1) The outer, rigid shell of the Earth, situated above the asthenosphere (a zone of the Earth's mantle that lies beneath the lithosphere and consists of several hundred kilometers of deformable rock) and containing the crust, continents and plates. (2) The solid outer layer of the Earth; includes both the land area and the land beneath the oceans and other water bodies.

Lithotroph. An organism that uses an inorganic substrate, such as ammonia (NH_3) or hydrogen (H), as an electron donor in energy metabolism. There are two types of lithotrophs: chemolithotrophs and photolithotrophs.

Litter. (1) Comprises the dead plant and animal debris or detritus on the soil surface. (2) Surface layer of the forest floor consisting of freshly fallen leaves, needles, twigs, stems, bark, and fruits. Usually denotes the freshly fallen plant material on the soil surface or ground.

Loam. A soil textural class that is 7% to 27% clay, 28% to 50% silt, and less than 52% sand. [also see *silt*; *sand*; *clay*; and *texture, soil*]

Loamy. Intermediate in texture and properties between fine-textured and coarse-textured soils. Includes all textural classes with the words loam or loamy as part of the class name, such as clay loam or loamy sand. [also see *texture, soil*]

Lodging. The collapse of top-heavy plants (particularly grain crops) due to excessive growth, adverse weather, and/or certain nutrient deficiencies.

Loess. (1) A fine, silty, unconsolidated accumulation of material transported and deposited by wind. (2) A widespread, loose deposit consisting mainly of silt. Most loess deposits formed during the Pleistocene as an accumulation of wind-blown dust carried from deserts, alluvial plains, or glacial deposits. [also see *eolian (soil material)*]

Lotic waters. Flowing waters, as in streams and rivers.

Low biuret urea ("LB" urea). Commonly defined as urea with less than 0.25% biuret (Carbamylurea $H_2NC(O)NHC(O)NH_2$), and commonly recommended for use in foliar applications. [also see *biuret*]

Luxury consumption (or uptake) [of plant nutrients]. (1) The absorption by plants of nutrients in excess of their need for growth. (2) The uptake by a plant of an essential nutrient in amounts exceeding what it needs. Luxury contents accumulated during early growth may be used for later growth. Example: if nitrogen is abundant in the soil, corn (*Zea mays*) may take in (or absorb) more than it requires.

Lysimeter. An apparatus installed in the soil for measuring percolation and leaching.

Lysis. The rupture of a cell, resulting in loss of cell contents.

Macrofauna. Soil animals that are greater than 1000 micrometers in length (e.g., vertebrates, earthworms, and large arthropods). [also see *microfauna; mesofauna;* and *microorganism, soil (microbe)*]

Macronutrient(s). A plant nutrient found at relatively high concentrations (usually greater than 500 milligrams per kilogram [ppm] in plants (Soil Science Society of America). Includes: carbon, hydrogen, oxygen, nitrogen, phosphorus, potassium, calcium, magnesium, and sulfur. [also see *plant nutrients, essential; primary plant nutrients; secondary nutrients;* and *micronutrient*]

Macroorganic matter. Organic fragments from any sources that are greater than 250 micrometers in size (generally less decomposed than humus).

Macropore. See *pore-size classification*.

Macroscopic. Objects or organisms that are large enough to be seen with the naked eye.

Mafic. Term used to describe the amount of dark-colored iron and magnesium minerals in an igneous rock.

Magma. Molten rock generated within the earth. Magma forms intrusive (solidifies below the surface) and extrusive (solidifies above the surface) igneous rocks.

Magnesium (Mg). Magnesium exists in soils and is absorbed by plants as the Mg^{2+} ion, where its oxidation state in the plant does not change. Magnesium is the central ion of chlorophyll and is actively involved in photosynthesis. Therefore, much of magnesium in plants is found in the chlorophyll. Seeds are also relatively high in magnesium though grain crops, such as corn, have low magnesium levels in the seed. Magnesium is mobile in plants and, for that reason, deficiency symptoms first appear on the older leaves as interveinal chlorosis.

Magnesium is also in very important in the base saturation of cations. <u>When present in quantities greater than approximately 10% of the base saturation, magnesium can significantly contribute to soil deflocculation</u>. Magnesium has the capacity to deflocculate soils to the degree or extent of approximately 10% of sodium's capacity to do the same. The addition of calcium can help rectify this problem. [also see *plant nutrients, essential; cation; base saturation;* and *deflocculation, soil*].

Magnesium sulfate. A fertilizer consisting chiefly of the chemical compound, magnesium sulfate, with or without combined water. Includes: Epsom salts ($MgSO_4 \cdot 7H_2O$), kieserite ($MgSO_4 \cdot H_2O$) and calcined kieserite ($MgSO_4$). [also see *calcination (calcining)*]

Management, soil. The sum total of all tillage and planting operations, cropping practices, fertilizer, lime, anhydrite and/or gypsum, irrigation, herbicide and/or insecticide applications, and other treatments conducted on or applied to a soil for the production of plants.

Manganese (Mn). Manganese is a metallic micronutrient existing in soils in several oxidation states of which the Mn^{2+} ion is the form most commonly absorbed by plants. Manganese functions primarily as a part of enzyme systems in plants. It activates several important metabolic reactions and plays a direct role in and for energy transfer during photosynthesis by aiding chlorophyll synthesis. Manganese accelerates germination and maturity, while increasing the availability of phosphorus and calcium.

Manganese occurs in soils in several oxidation state with Mn^{2+} and Mn^{4+} the predominant forms. The relative existence of these two forms is controlled by pH, redox potential (oxygen level of the soil), and organic matter interrelations. The lower the pH and oxygen content of the soil the greater the amount of Mn^{2+} in the soil. The availability of Mn^{2+} at pH <5.5 and with low soil oxygen levels can be sufficient to induce toxicity. Conversely, high pH, well-drained soils can result in manganese deficiencies.

Manganese is immobile in plants so any deficiency symptoms first appear as interveinal chlorosis on the younger leaves. [also see *interveinal chlorosis*; *plant nutrients, essential*; and *cation*]

Manganese sulfate (MnSO₄). A solid chemical compound used as a fertilizer source of manganese for plants with 26-28% manganese. Manganese sulfate is a moderately water and acid soluble manganese fertilizer. The term manganese sulfate when applied to an ingredient of a mixed fertilizer designates anhydrous manganese sulfate (MnSO₄) (AAPFCO). However, several hydrates, including MnSO₄·H₂O and MnSO₄·3H₂O, are also used as fertilizers.

Mantle. That portion of the interior of the Earth that lies between the crust and the core.

Manure. Any substance composed primarily of animal excrement, plant remains, or mixtures of those substances. Also, the excreta of animals, with or without an admixture of bedding or litter, fresh or at various stages of further decomposition or composting. In some countries manure denotes any fertilizer material. [also see *compost*].

Approximate Nutrient Composition of Various Animal Manures and Composts (fresh weight basis)					
Manure Type	Dry Matter	NH_4^+-N	Total N	P_2O_5	K_2O
	%	---------- lbs/ton ----------			
Swine, no bedding	18	6	10	9	8
Swine, with bedding	18	5	6	7	7
Beef, no bedding	52	7	21	14	23
Beef, with bedding	50	8	21	18	26
Dairy, no bedding	18	4	9	4	10
Dairy, with bedding	21	5	9	4	10

Sheep, no bedding	28	5	18	11	26
Sheep, with bedding	28	5	14	9	25
Poultry, no litter	45	26	33	48	34
Poultry, with litter	75	36	56	45	34
Turkey, no litter	22	17	27	20	17
Turkey, with litter	29	13	20	16	13
Horse, with bedding	46	4	14	4	14
Poultry compost	45	1	17	39	23
Dairy compost	45	<1	12	12	26
Mixed compost: Dairy/Swine/ Poultry	43	<1	11	11	10

University of Minnesota Extension. 2005.
Total N = ammonium-N plus organic N

Manure tea. A water extract of manure that usually contains soluble nutrients, relatively high levels of nitrates, salts, phosphorus, and potassium; and contains high numbers of bacteria (unless there has been an antibiotic used in the animal feed). Manure tea usually contains high numbers of protozoa, extremely low fungal biomass, and can contain high numbers of nematodes (often root-feeding nematodes). Manure tea almost always contains human and animal pathogens. [also see *compost tea*; *compost extract*; and *compost leachate*]

Marl. An earthy, unconsolidated deposit consisting chiefly of calcium carbonate mixed with clay in approximately equal proportions (35% to 65% of each); formed primarily under freshwater lacustrine conditions, but varieties associated with more saline environments also occur. [also see *lacustrine*]

Marsh. Periodically wet or continually flooded areas with the surface not deeply submerged. Covered dominantly with sedges, cattails, rushes, or other hydrophytic plants. [also see *hydrophytic (plants)*]

Mass flow (nutrient). Mass flow, or convection, is considered to be the most important mode of nutrient uptake. This mechanism relates to nutrient mobility with the movement of soil water towards the root surface where absorption through the roots takes place along with water. Some nutrient ions are called mobile nutrients while others which move only a few millimeters are called immobile nutrients. Mobile nutrient ions such as nitrate, chloride and sulfate, are not absorbed by the soil colloids and are mainly in solution. Such nutrient ions are absorbed by the roots along with soil water. Mass flow is responsible for supplying the root with much of the plant needs for nitrogen, calcium and magnesium, when present in high concentrations in the soil solution, but does not do so in the case of phosphorous or potassium. The nutrient uptake through mass flow is largely dependent on the moisture status of the soil and is highly influenced by the soil physical properties controlling the movement of soil water. [also see *diffusion;* and *root interception*]

Massive. Lack of soil structure in coherent materials; structureless but held together. Often associated with limited calcium and soil organic matter in the soil profile. [also see *anhydrite*; *gypsum;* and *organic matter, soil*]

Matric potential. See *water potential, soil*.

Matter. Anything that has mass and occupies space.

Mature compost. Composts that has been cured to a stabilized state, characterized by being rich in readily available forms of plant nutrients and low in readily available carbon compounds. Mature composts have carbon:nitrogen ratios less than 25:1. [also see *carbon:nitrogen ratio*]

Mature soil. A well developed soil; usually with clearly defined diagnostic horizons.

Maximum economic yield (MEY). (1) Yield at which unit costs of production are lowered to the point of highest net return per acre. (2) The most profitable yield. The MEY is generally achieved through implementation of Best Management Practices. [also see *Best Management Practices*]

Medium (plural media). Any liquid or solid material prepared for the growth, maintenance, or storage of microorganisms.

Meristem. Plant tissue capable of undergoing mitosis, and so giving rise to new cells and tissues, as at growing tips (e.g., apical meristem). The region of active cell division in plants. The new cells formed become modified to form the various tissues of the plant, e.g., the epidermis and cortex. [also see *epidermis*; and *cortex*]

Mesofauna. Small soil organisms such as nematodes, insect larvae, small arthropods, worms, and insects. Mesofauna range in size from 200 micrometers to 1 centimeter in length. [also see *microfauna*; *macrofauna;* and *microorganism, soil (microbe)*]

Mesophilic. Thriving at moderate temperatures [between 20 °C and 40 °C (68 °F to 104 °F)].

Mesophilic bacteria. Bacteria that thrive at moderate temperatures [between 20 °C and 40 °C (68 °F to 104 °F)].

Mesophilic stage of decomposition. A stage in the composting process characterized by bacteria that are active in the moderate range of temperatures [between 20 °C and 40 °C (68 °F to 104 °F)]. The mesophilic stage is also associated with a moderate decomposition rate.

Mesosphere. In the atmosphere, the region immediately above the stratosphere and immediately below the thermosphere. The mesosphere begins about 50 kilometers (approximately 31 miles) high at the stratopause and ends about 80 kilometers (approximately 50 miles) high at the mesopause. The top of the mesosphere, called the mesopause, is the coldest place on Earth. Temperatures in the upper mesosphere fall as low as −100 °C (173 K; −148 °F), varying according to latitude and season. [also see *atmosphere*; *stratosphere;* and *thermosphere*]

Metabolism. The sum of all chemical processes and reactions occurring within a living cell or organism. These include catabolism, the energy-releasing breakdown of molecules; and anabolism, the synthesis of complex molecules and new protoplasm. [also see *catabolism;* and *anabolism*]

Metal. A substance that conducts heat and electricity, is shiny and reflects many colors of light, and can be hammered into sheets or drawn into wire. Metals lose electrons easily to form cations. About 80% of the known chemical elements are metals. [also see *non-metal*]

Metamorphosis. A process of developmental change whereby a larva reaches adulthood only after a drastic change in morphology; occurs in most amphibians and insects, for some insects, this change may include another stage (pupa) before the adult stage. [verb: metamorphose]. [also see *larva;* and *pupa*]

Metamorphic rock. Igneous or sedimentary rock that has changed because of high temperature, high pressure, and/or the chemical environment while deep in the crust of the Earth. [Example: marble is metamorphosed limestone ($CaCO_3$)]. [also see *igneous;* and *sedimentary*]

Methane (CH_4). An odorless, colorless, flammable gas. Methane is the simplest of all hydrocarbons with a formula of CH_4. Methane is released naturally into the air from organic decomposition in anaerobic environments such as marshes, swamps, landfills, and rice fields, and from symbiotic microbes in the guts of ruminant animals (such as cattle and sheep), and sewage sludge. Methane is released from methane producing bacteria (methanogens) that live in these anaerobic places. Methanogens in termite guts are the source of methane released by termites. [also see *anaerobic; decomposition;* and *hydrocarbon*]

Methanogenesis. The biological production of methane.

Microaggregate. Clustering of clay packets stabilized by organic matter and precipitated inorganic materials.

Microbe, soil. See *microorganism, soil.*

Microbial/microbiology. Pertaining to the study of microorganisms.

Microbial biomass. The total mass of living microorganisms in a given volume of mass of soil. The total weight of all microorganisms in a particular environment.

Microbial population. The sum of living microorganisms in a given volume or mass of soil.

Microclimate. The climate of a very small region. [Example: a small creek or stream valley]

Microfauna. Small animals of microscopic size including protozoa, nematodes, and arthropods generally less than 200 micrometers long. [also see *macrofauna; mesofauna;* and *microorganism, soil (microbe)*]

Microflora. Small plants of microscopic size including bacteria, actinomycetes, fungi, algae, and viruses.

Microhabitat. Clusters of microaggregates with associated water within which microbes function. May be composed of several microsites (e.g., aerobic and anaerobic).

Micrometer. One-millionth of a meter, or 10^{-6} meter, the unit usually used for measuring microorganisms.

Micron. A unit of length equivalent to 39.37 millionths of an inch, or 0.001 millimeter. One millimeter equals 1000 microns. A useful unit in dealing with particle size. Example: a 200-mesh sieve has a screen opening of 74 microns.

Micronutrient(s). (1) A plant nutrient found in relatively small amounts [usually less than 100 milligrams per kilogram (ppm)] in plants [Soil Science Society of America]. (2) Inorganic chemical elements required only in very small, or trace, amounts for plant growth. These are usually boron, chlorine, copper, iron, manganese, molybdenum, zinc and sometimes cobalt and nickel. [also see *plant nutrients, essential*; *macronutrient(s);* and *secondary nutrients*]

Microorganism, soil (microbe). Organisms that live in the microscale environments within and between soil particles, too small to be seen with the naked eye (< 0.1 millimeter). Differences over short distances in pH, moisture, pore size, and the types of food available create a broad range of habitats and organisms themselves. Soil microorganisms are typically cysts, amoebas, flagellates, bacterial colonies, nematodes, ciliates, fungal hyphae and spores, and actinomycete hyphae and spores.

Microorganisms occur wherever organic matter occurs, but mostly in the top few inches of the soil. However, some microorganisms have been found as deep as 10 miles (16 kilometers) in oil wells. [also see *macroorganism*; *microflora;* and *microfauna*] Note: Microbe is an imprecise term referring to any microscopic organism.

Micropore. See *pore-size classification.*

Microrelief. Small differences in relief that have differences in elevation up to approximately 2 meters (approximately 1.63 yards).

Microscopic. Objects or organisms that are too small to be seen with the naked eye.

Microsite. A small volume of soil where biological or chemical processes differ from those of the soil as a whole; such as an anaerobic microsite of a soil aggregate or the surface of decaying organic residues.

Milligrams per liter (mg L^{-1}). A unit of the concentration of a constituent in water or wastewater. It represents 0.001 gram of a constituent in 1 liter of water. One mg L^{-1} is approximately equal to one part per million (ppm).

Milliequivalent (or milligram equivalent) [meq]. 1/1000 (or milligram equivalent) of an equivalent weight.

Milliequivalents per liter (meq L^{-1}). The most meaningful method of reporting the major chemical components of irrigation water. Salts are the combinations of the cations sodium, calcium, magnesium and potassium, etc., and the anions chloride, sulfate, and bicarbonate, etc. in definite weight ratios. These weight ratios are based upon the atomic weight of each constituent and upon the valence or electrical charge of each. An equivalent weight of an ion is the atomic weight divided by its valence for equivalent weights of the common ions. Therefore, meq L^{-1} is a measurement of charge concentration per liter of water. [Note: 1 milliequivalent (or milligram equivalent) = 1/1,000 of an equivalent. [Example: in the case of sodium chloride, 1 meq L^{-1} would be 23 milligrams of sodium and 35.5 milligrams of chloride in 1 liter of water]

Equivalent Weights of Common Water Chemicals

Ion Name	Symbol	Equivalent Wt	Atomic Wt
Calcium	Ca^{2+}	20.04	40.08
Magnesium	Mg^{2+}	12.15	24.30
Sodium	Na+	22.99	22.99

Potassium	K+	39.10	39.10
Bicarbonate	HCO_3^-	61.02	61.02
Carbonate	CO_3^{2-}	30.00	60.01
Sulfate	SO_4^{2-}	48.03	96.06

Note: ppm = equivalent weight X milliequivalents per liter (meq L^{-1}); milliequivalents per liter (meq L^{-1}) = ppm ÷ equivalent weight

Compiled From Various Sources

Example: An irrigation water analysis reports 521 parts per million sodium and 122 parts per million bicarbonate:

521 ÷ 22.99 = 22.66 meq L^{-1} sodium; and
122 ÷ 61.02 ≈ 2.0 meq L^{-1} bicarbonate in the irrigation water

Mineral. (1) A naturally occurring, homogeneous, inorganically formed, solid, chemical element, with a definite composition and an ordered atomic arrangement. (2) Any mineral that occurs as a part of or in the soil. (3) A natural inorganic compound with definite physical, chemical, and crystalline properties (within the limits of isomorphism), that occurs in the soil. [Examples: anhydrite ($CaSO_4$), gypsum ($CaSO_4 \cdot 2H_2O$), calcite ($CaCO_3$), and dolomite ($CaMg(CO_3)_2$). [compare to *rock*]

Mineral, agricultural. Any substance with nitrogen (N), available phosphoric acid (P_2O_5), or soluble potash (K_2O), singly or in combination, in amounts less than 5%, which is distributed for farm use, or any substance only containing recognized essential secondary nutrients or micronutrients in amounts equal to or greater than the minimum amount specified by regulations, and distributed in this state as a source of these nutrients for the purpose of promoting plant growth. This category includes, but is not limited to: gypsum, liming materials, manure, wood fly ash, and sewage sludge not qualifying as commercial fertilizer. [also see *fertilizer, commercial*]

Mineral soil. A soil consisting predominantly of, and having its properties determined mainly by, mineral (inorganic) matter. Mineral soils usually contain less than 20% organic matter, but may contain an organic surface layer up to 30 cm (approximately 11.8 inches) thick. [also see *organic soil*]

Mineralizable. The term 'mineralizable' is most commonly used in reporting nitrogen (and to a lesser degree carbon). It refers to the amount of N that is 'mineralized' or made available by microorganisms as they decompose organic materials. Researchers usually arrive at this number using either lab or field (*in situ*) incubations of soil over a given period of time. Methods can vary from one study to the next, and must be kept in mind when interpreting results.

Mineralization. The conversion of an element from an organic form to an inorganic state as a result of microbial activity. [Example: the conversion of organic nitrogen in amino groups ($-NH_2$) to the inorganic ammonium (NH_4^+) form]

Minor elements. See *micronutrient*.

Mites. Very small members of the phylum arthropod, which includes spiders. Mites occur in large numbers in many organic surface soils.

Mixed fertilizers. See *fertilizer, mixed*.

Mixotroph. Organism able to assimilate organic compounds as carbon sources while using inorganic compounds as electron donors. [also see *autotroph*; and *heterotroph*]

Moder. A type and kind of decomposition and soil organic matter formation which reproduces advanced but incomplete humification of the remains of organisms due to lack of good aeration.

Mold. Fungi that grow in the form of multicellular filaments called hyphae. In contrast, microscopic fungi that grow as single cells are called yeasts. A connected network of these tubular branching hyphae has multiple, genetically identical nuclei and is considered a single organism; referred to as a colony or in more technical terms a mycelium. [also see *yeast; hyphae;* and *mycelium*]

Mole (mol). (1) The quantity of a given substance that contains as many molecules as the number of atoms in exactly 12 grams of carbon-12. (2) The amount of substance containing Avogadro's number (6.02×10^{23}) of molecules. (3) One mole of substance is its molecular mass in grams. For Example: 1 mole of carbon-

12 is 12 grams; and 1 mole of table salt, NaCl, is 23 + 35.5 = 58.5 grams.

Molecule. (1) Result of two or more atoms combining by chemical bonding. (2) A definite group of atoms that are chemically bonded together, this is, tightly connected by attractive forces. Molecules are "super blocks" made up of atoms. [Examples: $CaSO_4 \cdot 2H_2O$ is the molecule gypsum; NH_3 is the molecule ammonia].

Molecular weight. (1) The sum total of the atomic weights of the atoms contained in the molecule. (2) The sum of the relative weights of the atoms in a molecule.

Example: Anhydrous ammonia (NH_3)	
Atoms	Atomic Weight
1 nitrogen atom	14.0067
3 hydrogen atoms [1.0079 X 3]	3.0237
molecular weight of NH_3 =	17.0304

Molybdenum (Mo). Molybdenum, a metallic micronutrient, is absorbed primarily as the molybdate anion MoO_4^{2-}, but also as $HMoO_4^-$ (bimolybdate). Plants require molybdenum in the smallest quantities of all the essential elements. Molybdenum is required for the synthesis and activity of the enzyme nitrate reductase. This enzyme system reduces NO_3^--nitrogen to NH_4^+-nitrogen in the plant. Molybdenum is also vital for the process of symbiotic nitrogen fixation by *Rhizobia* bacteria in legume root nodules. A deficiency can cause nitrogen deficiency symptoms in legume crops since symbiotic soil bacteria must have Mo^{2+} to fix nitrogen from the atmosphere. Molybdenum availability increases as soil pH increases, the opposite of most micronutrients. Therefore, deficiencies are more likely to occur in acid soils. [also see *plant nutrients, essential;* and *anion*].

Mohs' scale of hardness. Relative hardness of minerals ranging from a rating of 1 for the softest (talc) to 10 for the hardest (diamond). [Examples: gypsum = 2 (soft), dolomite = 3.5 to 4 (semi-hard), pyrite (iron sulfide, FeS_2) = 6 to 6.5 (hard)].

Monoammonium phosphate (MAP) (10-53-0 to 12-61-0) ($NH_4H_2PO_4$). A water-soluble fertilizer composed of ammonium phosphates, principally monoammonium phosphate, resulting from the ammoniation of phosphoric acid. The guaranteed percentage of nitrogen and available phosphate should be stated as part of the name.

Monocalcium phosphate. ($Ca(H_2PO_4)_2$). Monocalcium phosphate, a calcium phosphate fertilizer, is a chemical compound with the formula $Ca(H_2PO_4)_2$; but is commonly found as the monohydrate, $Ca(H_2PO_4)_2 \cdot H_2O$, also known as super phosphate. [also see *super phosphate;* and *triple superphosphate*]

$$3\, Ca_3(PO_4)_2 + 6\, H_2SO_4 \longrightarrow 6\, CaSO_4 + 3\, Ca(H_2PO_4)_2$$

Monocotyledons (also known as monocots). Monocotyledons are one of two major groups of flowering plants that are traditionally recognized, the other being dicotyledons, or dicots; characterized by having a single cotyledon, floral organs arranged in threes or multiples of three, and parallel-veined leaves. Monocots include grasses, cattails, lilies, and palm trees. [also see *dicotyledons (also known as dicots)*]

Monosaccharide. A sugar not decomposable to simpler sugars by hydrolysis (the chemical reaction in which a chemical compound decomposes by reaction with water). They are the simplest form of sugar and are usually colorless, water-soluble, crystalline solids. Some monosaccharides have a sweet taste. Examples of monosaccharides include glucose (dextrose), fructose (levulose), galactose, xylose and ribose. Monosaccharides are the building blocks of disaccharides (such as sucrose) and polysaccharides (such as cellulose and starch). [also see *hydrolysis; disaccharide; glucose;* and *polysaccharide*]

Montmorillonite. A very soft, hydrous phyllosilicate group of minerals that typically form in microscopic crystals, forming a clay. It is named after Montmorillon in France. Montmorillonite, a member of the smectite family, is a 2:1 clay, meaning that it has 2 silicon tetrahedral sheets sandwiching a central aluminum octahedral sheet. It is the main constituent of the volcanic ash weathering product bentonite. Considerable expansion of montmorillonite may be caused by water. [also see *bentonite;* and *kaolinite*]

Moraine. Any type of constructional topographic form consisting of till, and resulting from glacial deposition.

Morphology. The form and structure of anything, usually applied to the shapes, parts, and arrangement of features in living and fossil organisms.

Mottling (mottles). Irregular soil mass spots or blotches of different colors or shades of colors interspersed with the dominant color. A common cause of mottling is impeded drainage.

Mor. Acid humus of cold welt soils which inhibits action of soil organisms and may form peat. [also see *mull*]

Muck soil. (1) A soil containing 20% to 50% organic matter. (2) An organic soil whose original organic plant part material is too decomposed to be recognizable. Muck soils contain more mineral matter and are usually darker in color than peat soils. [also see *peat soil*]

Mulch. Material applied to or left on the soil surface in order to prevent rapid changes in soil temperature, to prevent erosion, to suppress weed growth, to add organic matter to the soil, and for decorative purposes. The main value of mulches is to reduce loss of water through evaporation, help control weeds, and reduce soil erosion. Mulching materials, besides crop residues, include sawdust, leaves, grass clippings, composts, etc., as well as paper and plastic specially prepared for mulching purposes.

Mulch, stubble. Stubble mulching is one of the early erosion control practices in which only part of the crop residue is incorporated somewhat into the soil. Most of the residue is left anchored in the soil but exposed at the soil surface.

Mulching. The application of a layer of material (e.g., composts and/or manures) to the surface of the soil, thus creating an interface that accepts water readily yet resists moisture loss through evapotranspiration.

Mulching, straw. The use of straw to create a surface mulch on all or part of the soil surface for soil or water conservation, for soil temperature management, and/or for weed suppression.

Mull. Humus of well-aerated moist soils formed by action of soil organisms (e.g., earthworms) on plant debris and detritus, and supporting plant growth. [also see *mor*]

Municipal organic materials. Organic materials generated by residential, business, institutional, and agricultural sources, which are then collected and sent to city and county waste facilities.

Municipal solid waste. Garbage. Refuse offering the potential for energy recovery; includes residential, commercial, and institutional wastes.

Muriate of potash (KCl). Muriate of potash (commercial potassium chloride) is defined as a fertilizer that contains 48% to 62% soluble potash (K_2O), chiefly as chloride.

Mushroom. Any of various fleshy fungi of the class Basidiomycota characteristically having an umbrella-shaped cap borne on a stalk; especially any of the edible kinds as those of the genus *Agaricus*.

Mushroom compost (soil). Commercial mushrooms are grown in a specially formulated and processed compost made from any combination of: wheat, oats or rye straw, peat moss, used horse manure and bedding straw, chicken manure, cottonseed or canola meal, dried blood, grape crushings from wineries, soybean meal, ground chalk, fertilizers (e.g., ammonium nitrate, urea, potash, etc.), calcium sulfate (anhydrite or gypsum), agricultural lime, cottonseed hulls, corncobs, cocoa bean shells, and/or chicken litter. Mushrooms generally can only be grown in this mixture one time and then compost has to be removed because the growing environment is very dark and moist and is subject to invasion by other fungi and bacteria. The compost is then removed and the residual spent compost waste product is then sold under various names as: (1) mushroom compost, (2) mushroom soil, (3) spent mushroom compost, and/or (4) spent mushroom substrate (SMS). The spent mushroom medium/mushroom compost is excellent for agriculture and horticulture as a good source of composting medium, organic matter, and soil amendment.

Mutualism. The interaction between organisms where both organisms benefit from the association. [also see *symbiosis*]

Mutualists. Two organisms living in an association that is beneficial to both, such as the association of roots with mycorrhizal fungi or with nitrogen fixing bacteria. [also see *symbiosis*]

Mycelium (plural mycelia). (1) A mass of hyphae that form the vegetative body of many fungal organisms. (2) The body of a fungus composed of many threads of tissue. Mushrooms do not reproduce by seed, but by spores. The spores germinate to produce threadlike structures known as hyphae. Collectively, a mass of hyphae are known as the mycelium. [also see *hyphae;* and *mold*]

Mycophagous. Organisms that consume fungi, such as mycophagous nematodes.

Mycorrhiza (plural mycorrhizae). Literally "fungus root." (1) Occurs when a fungus (basidiomycete or zygomycete) weaves around or into a plant's roots and forms a symbiotic relationship. (2) The association, usually symbiotic, of specific fungi with the roots of higher plants. Fungal hyphae absorb minerals from the soil and pass them on to the plant roots while the fungus obtains carbohydrates from the plant. [also see *ectomycorrhiza; endomycorrhiza; symbiosis;* and *vesicular arbuscular*]

Mycorrhizosphere. The unique microbial community that forms around a mycorrhiza.

Nanopore. Soil pore having dimensions measured in nanometers (nm) [one billionth of a meter (m^{-9})]. Materials encased in nanopores are beyond the reach of microorganisms and enzymes.

Native plants. The plants of a given ecosystem that grew prior to European contact. Native plants have co-evolved with animals, fungi, and microorganisms to form a complex network of relationships. These plants are the foundation of native ecosystems, or natural communities.

Natural erosion. See *erosion, natural*.

Necroses. Plant diseases characterized by areas of dead tissue (necrosis) and mosaic diseases, in which the leaves or other plant parts display mottling generally caused by plant viruses.

Necrosis. (1) The appearance of dead parts of plants due to a lack of plant growth factors or the presence of diseases or toxins. (2) Death associated with discoloration and dehydration of all or parts of plant organs, such as leaves. (3) Dead plant tissue. (4) The gradual decay of trees or plants. Necrosis can also be confused with the normal senescence of plant parts. [also see *senescence;* and *chlorosis*].

Nematode. Generally small (0.5 to 1.5 millimeters long), elongated, cylindrical, non-segmented worms. Some are parasitic on plant roots; others parasitize insects or fungi (see mycophagous nematodes). Also called nematode worm and roundworm. Most nematodes live free in the soil. [Examples: root knot, ring, and lesion nematodes; all are considered harmful to grapes and other crops grown in California and the west].

Neutral, soil. A soil in which the surface layer, at least to normal plow depth, is neither acid nor alkaline in reaction. Put into practice, this means the soil is within the pH range of 6.6 to 7.3. [also see *pH, soil; acid soil;* and *alkaline soil*]

Neutralism. A lack of interaction between two organisms in the same habitat.

Niche. The functional role of a given organism within its habitat.

Nickel (Ni). While not universally accepted as an essential plant nutrient, nickel may possibly be essential for certain types of nitrogen nutrition and certain plant families. Nickel is certainly beneficial for legumes. Nickel, as Ni^{2+}, is the metal component of the enzyme urease that catalyzes the reaction:

$$CO(NH_2)_2 \text{ (urea)} + H_2O \xrightarrow{\text{urease}} 2NH_3 \text{ (ammonia)} + CO_2$$

Therefore, nickel is essential for plants supplied with urea and for those in which ureides (any of the derivatives of urea) are important in nitrogen metabolism. High levels of nickel have been shown to induce zinc and/or iron deficiency because of cation competition. [also see *cation; plant nutrients, essential;* and *ureide*]

Nitrate reduction (biological). The process whereby nitrate (NO_3^-) is reduced by plants and microorganisms to ammonium (NH_4^+) for cell synthesis (nitrate assimilation, assimilatory nitrate reduction) or to various lower oxidation states (N_2, N_2O, NO) by bacteria using nitrate as the terminal electron acceptor in anaerobic respiration. Nitrate reduction occurs in anaerobic conditions. The "bugs" that eat it don't like oxygen.

Nitrification. (1) The bacterial oxidation of ammonium salts to nitrites (via *Nitrosomonas*) and further oxidation of nitrites to nitrates (via *Nitrobacter*). (2) Biological oxidation of ammonium to nitrite and nitrate, or a biologically induced increase in the oxidation state of nitrogen. The nitrifying organisms achieve maximum potential under the following conditions: (a) an abundance of proteins to release ammonium, (b) adequate aeration, (c) a moist but not overly wet soil, (d) a large amount of calcium, and (e) optimum temperature between 20 °C and 40 °C (68 °F to 104 °F). [also see *autotrophic nitrification;* and *denitrification*]

Note: For each pound of nitrogen as ammonium or forming ammonium from urea, ammonium nitrate, and anhydrous ammonia, it takes approximately 1.8 pounds of pure calcium carbonate ($CaCO_3$) to neutralize the residual acidity. [also see *acid soil (soil acidity), residual*]

The chemical formulas for nitrification:

$$2NH_4^+ + 3O_2 \xrightarrow{\text{Nitrosomonas}} 2NO_2^- + 2H_2O + 4H^+ + \text{energy}$$

$$2NO_2^- + O_2 \xrightarrow{\text{Nitrobacter}} 2NO_3^- + \text{energy}$$

Nitrobacter. Bacteria that convert nitrites (NO_2^-) into nitrates (NO_3^-) in the nitrification process. [also see *Nitrosomonas;* and *nitrification*]

Nitrogen (N). (1) N_2, a colorless, tasteless, odorless nearly unreactive diatomic gas molecule which makes up 78.08% of the atmosphere. Atmospheric nitrogen (N_2) is converted by nitrogen fixation and nitrification into compounds used by plants and animals. (2) Nitrogen is the only plant nutrient that is absorbed by plant roots in both an anion (NO_3^-) and cation (NH_4^+) form. Nitrogen is an essential nutrient and constituent of every living cell, plant or animal. In plants, it is part of the chlorophyll molecule (and, therefore, photosynthesis), amino acids, proteins, and many other compounds. Lack of adequate nitrogen and chlorophyll diminishes plant utilization of sunlight as an energy source, decreases production of carbohydrates, and limits essential functions such as nutrient uptake and protein syntheses. As a result, growth is stunted, yields are reduced, and crop quality is impaired.

Nitrogen is very mobile within the plant and moves from older leaves to new growth areas under deficiency conditions, producing leaf chlorosis (yellowing) beginning with the older leaves. As the deficiency becomes more acute, the chlorosis can extend over the entire plant. [also see *anion; cation; chlorosis; nitrification;* and *plant nutrients, essential*]

Nitrogen cycle. (1) The circulation of the element nitrogen (N) in the biosphere, from nitrogen fixation to the release of free nitrogen by denitrifying bacteria. Nitrogen is present in the environment in a wide variety of chemical forms including organic nitrogen, ammonium (NH_4^+), nitrate (NO_3^-), and the nearly unreactive diatomic molecule: nitrogen gas (N_2). (2) The processes of the nitrogen cycle transform nitrogen from one chemical form to another. Many of the processes are carried out by microbes either to produce energy or to accumulate nitrogen in the form needed for growth.

Nitrogen fixation. See *dinitrogen fixation*.

Nitrosomonas. Bacteria that convert ammonium (NH_4^+) into nitrites (NO_2^-) in the nitrification process. [also see *Nitrobacter;* and *nitrification*]

Nitrous oxide (N_2O). Nitrous oxide is a by-product of biological activity of a symbiotic bacteria living (i.e., *Rhizobia*) in leguminous plant roots. It is a principal greenhouse gas, and is also "laughing gas" used in medicine as a gentle general anesthetic.

Nodule, bacterial. (1) Enlargements or swellings on the roots of legumes and certain other plants inhabited by symbiotic nitrogen-fixing bacteria. (2) A growth developed on the root of plants, especially legumes, in response to the stimulus of root nodule bacteria or actinomycetes. [also see *Rhizobia; symbiotic bacteria; leghemoglobin;* and *legumes*]

Nonacid-forming fertilizer. A fertilizer that is not capable of increasing the residual acidity of the soil. An example is calcium ammonium nitrate (CAN). [opposite = acid-forming fertilizer]

Nonmetal. A nonmetal is a substance that conducts heat and electricity poorly, is brittle or waxy or gaseous, and cannot be hammered into sheets or drawn into wire. Nonmetals gain electrons easily to form anions. About 20% of the known chemical elements are nonmetals (the rest are metals). [also see *metal*]

Non-point source pollution. Pollution discharged over a wide land area, not from one specific location. These are forms of diffuse pollution caused by sediment, nutrients, organic and toxic substances originating from land-use activities which are carried to lakes and streams by surface runoff. Non-point source pollution is contamination that occurs when rainwater, snowmelt, or irrigation washes off fields, city streets, or suburban backyards. As this runoff moves across the land surface, it picks up soil particles and pollutants, such as nutrients and pesticides. Non-point source pollution is difficult to pinpoint physically but can be classified by type: urban runoff, agriculture, mining, septic tank leach fields, and silviculture. [also see *point source pollution;* and *silviculture*]

Nucleus (atom). An atom's core and contains protons and one or more neutrons (except hydrogen, which has no neutrons).

Nucleus (cell). Membrane-enclosed structure containing the genetic material (DNA) organized in chromosomes.

Nutrient, plant. See *plant nutrients, essential.*

Nutrient balance. A ratio among concentration of nutrients essential for plant growth which permits maximum growth rate and yield.

Nutrient deficiency. A low concentration of an essential element that reduces plant growth and prevents completion of the normal plant life cycle. Nutrient deficiency symptoms can easily be confused with toxicity symptoms.

Nutrient uptake. The process of plant absorption of nutrients, usually through the roots. Small amounts of some, but not all, nutrients may be absorbed through the leaves following foliar application. Root nutrient uptake is affected by expendable energy supplies in the plant. Other factors that affect nutrient uptake include soil temperature, soil aeration, soil moisture, soil structure, soil pH, concentrations of nutrients in the soil, interactions of various nutrients and plant rooting patterns; plus an extensive combination of other factors. [also see *diffusion; mass flow;* and *root interception*]

Nutrient stress. A condition occurring when the quantity of nutrient, or nutrients, available reduces growth. Nutrient stress can be from either a deficient or toxic concentration.

Nutrients, available (vs. total). Available nutrients are exactly that – available to plants. When the 'total' amount of a nutrient is reported, it represents all that is readily available in addition to that which is 'tied-up'. Total nutrients are significant since some portion eventually becomes available as decomposition or soil mineral weathering proceeds. Nitrogen is an important working example. Total N in the top 40 cm of mineral soil might be 1500 kg ha^{-1}. At the same time, only about 30 kg ha^{-1} of this might be found in the available form. In this case, knowing the total amount of N in the soil is key as this is the pool from which the available form of N will come from.

Nymph. In aquatic insects, the larval stage.

"O" horizon, soil. The "O" stands for "organic," given that this surface layer/horizon is dominated by the presence of large amounts of organic material in varying stages of decomposition. The O horizon should be held distinct from the layer of leaf litter covering many heavily vegetated areas, as these contain no weathered mineral particles and are, therefore, not part of the soil itself.

Obligate. (1) Adjective referring to an environmental factor that is always required for growth (for example, oxygen). (2) An organism that can grow and reproduce only by obtaining carbon and other nutrients from a living host, such as obligate symbionts (e.g., many lichens). [also see *obligate symbiont*; *facultative symbiont; and symbiont*]

Obligate (aerobic organisms) aerobes. An organism that can only grow and is only metabolically active in the presence of oxygen.

Obligate (anaerobic organisms) anaerobes. An organism that can only grow and is only metabolically active in the absence of oxygen.

Obligate symbiont. Some symbiotic relationships are obligate, meaning that both symbionts entirely depend on each other for survival. For example, many lichens consist of fungal and photosynthetic symbionts (e.g., algae) that cannot live on their own. Others are facultative, meaning that they can but do not have to live with the other organism. [also see *facultative symbiont; and symbiont*]

Oligotroph. A microorganism specifically adapted to grow under low nutrient supply. Thought to subsist on the more resistant soil organic matter and be little affected by the addition of fresh organic materials. Sometimes a synonym for autochthonous.

Omnivore. Literally, an organism that will eat anything. Refers to animals that do not restrict their diet to just plants or other animals.

Order, soil. The highest level of soil classification. There are 11 soil orders: (1) Entisols, (2) Inceptisols, (3) Spodosols, (4) Ultisols, (5) Alfisols, (6) Vertisols, (7) Oxisols, (8) Histosols, (9) Andisols, (10) Aridosols, and (11) Mollisols.

Ore. A rock or mineral from which a metal can be economically produced.

Organic. Pertaining to living organisms in general, to compounds formed by living organisms, and to the chemistry of compounds containing carbon. Not to be confused with "certified organic." [also see *certified organic; fertilizer; inorganic (compounds); natural organic;* and *organics*]

Organic carbon. Carbon content is commonly used to characterize the amount of organic matter in soils. The Walkley-Black procedure is generally used to determine <u>oxidizable organic carbon</u>. That number is then multiplied by 1.724 to get <u>oxidizable organic matter</u>. The values for oxidizable organic carbon and oxidizable organic matter are then multiplied by 1.30 to obtain total organic carbon and total organic matter, respectively.

Organic matter, soil. The important soil fraction composed of carbonaceous material of plant and animal origin (exclusive of undecayed plant and animal residues) containing essential plant nutrients. Organic matter provides adhesives for soil particles, and is characterized by a high cation exchange capacity and absorptive capacity for the soil solution.

Above ground plants (the phytomass) are generally excluded from discussions of soil organic matter, but living roots are generally included. Soil organic matter is a significant part of the soil's ability to maintain structure, retain air, water and nutrients, and stimulate microbial activity. Loss of organic matter leads to low fertility, low water holding capacity, compaction, increased erosion, and diminished productive capacity. Soil organic matter is commonly determined as the amount of organic material contained in a soil sample passed through a 2 millimeter sieve. [also see *humus*]

Organic residue, soil. Animal and vegetative materials added to the soil of recognizable origin.

Organic soil. (1) A soil that contains at least 20% organic matter (by weight) if the clay content is low and at least 30% if the clay content is as high as 60%. (2) A soil in which the sum of the thicknesses of layers containing organic soil materials is generally greater than the sum of the thicknesses of mineral layers. [also see *mineral soil*]

Organics. Generally speaking, "organics" refers to materials that are or were once living, such as leaves, grass, agricultural crop residues, or food scraps. Nature has a way of decomposing these organic materials and naturally reusing them. When humans are involved, the decomposition process is called composting, and the resultant material is compost. [also see *compost*; *organic*; and *fertilizer, organic*]

Osmosis. (1) The movement of water molecules through a thin membrane. (2) Diffusion of water through a membrane from a region of low solute concentration to one of higher concentration. The osmosis process occurs in plants and animals, and is also one method of desalinizing saline water.

Osmotic potential. See *water potential, soil*.

Osmotic pressure. Pressure exerted in living bodies as a result of unequal concentrations of salts on both sides of a cell wall or membrane. Water moves from the area having the lower salt concentration through the membrane into the area having the higher salt concentration and, therefore, exerts additional pressure on the side with higher salt concentration.

Outcrop. (1) That part of a geologic formation or structure that appears at the surface of the Earth. (2) An actual exposure of bedrock at or above the ground surface.

Outwash. Glacially deposited soil parent material worked and graded by water action from melting glacial ice.

Overburden. (1) The upper part of a sedimentary deposit, compressing and consolidating the materials below. (2) The loose soil or other unconsolidated material overlying bedrock, either transported or formed in place.

Oxamide (fertilizer grade). A fertilizer that is the diamide of oxalic acid of the formula $C_2H_4N_2O_2$ which contains 28% to 32% nitrogen. Oxamide is a source of slowly available nitrogen.

Oxic. Containing oxygen or aerobic. Usually used in reference to a microbial habitat.

Oxidants. An oxidizing agent (also called an oxidant or oxidizer) can be defined as either: (1) a chemical compound that readily transfers oxygen atoms, or (2) a substance that gains electrons in a redox chemical reaction. [also see *redox*]

Oxidation. Combination with oxygen. The loss of one or more electrons by an ion or molecule. [also reduction and redox]. Biological oxidation is the process by which living organisms, in the presence of oxygen, convert organic matter into a more stable or a mineral form. [also see *reduction;* and *redox*]

Oxidation-reduction (redox) reaction. See *redox*.

Example 1: Oxidation of ferrous (reduced) iron to ferric (oxidized) iron:

$$2\ Fe^{2+} + H_2O_2 + 2\ H^+ \longrightarrow 2\ Fe^{3+} + 2\ H_2O$$

Example 2: Oxidation of elemental iron to iron(III) oxide by oxygen (commonly known as rusting):

$$4\ Fe + 3\ O_2 \longrightarrow 2\ Fe_2O_3$$

Oxygen (O). Oxygen (O) is the most abundant and most widely distributed element in nature, comprising about 46.6% of the Earth's surface. All major classes of structural molecules in living organisms, such as proteins, carbohydrates, and fats, contain oxygen, as do the major inorganic compounds that comprise animal shells, teeth, and bone. Oxygen in the form of O_2 is produced from water by cyanobacteria, algae and plants during photosynthesis and is used in cellular respiration for all complex life. [see *oxygen [molecular, (O_2)]*]

Oxygen [molecular, (O_2)]. Oxygen found on Earth as a gas constitutes about 20.95% of the air we breathe. Elemental molecular oxygen consists of two oxygen atoms bonded together, O_2, and has no color, odor, or taste. Molecular oxygen is present in both the atmosphere and dissolved in the oceans and freshwater sources exposed to the atmosphere. Oxygen (O_2), an essential plant nutrient, is found in aerobic soils in about 20.3% by volume as compared with that of 20.95% in the atmosphere. By comparison, the CO_2 level of the atmosphere by volume is about 0.03% whereas in the soil it is higher in the order of 0.2 to 1%.

Higher CO_2 and lower O_2 levels result from the respiration of living organisms in which O_2 is consumed and CO_2 is released. The respiration of plant roots depends to a high extent on the O_2 supply in the soil air. Respiration provides the energy for various metabolic processes including active ion uptake by plant roots. Oxygen supply to roots and other aerobic organisms in the soil not only depend on the O_2 content of the soil air, but also on the total volume of air present in the soil. Roots deprived of oxygen causes most land plants to die within a short time. The quantity of soil air declines as the water content of the soil increases since air, which normally fills the larger soil pores, is replaced by water. Therefore, an increase in soil water content depresses aerobic processes and supports anaerobic processes. [also see *plant nutrients, essential;* and *respiration*]

Oxygen demand. The need for molecular oxygen to meet the needs of biological and chemical processes in water. Even though very little oxygen will dissolve in water, it is extremely important in biological and chemical processes.

Ozone (O_3). A molecule that consists of three oxygen atoms bonded together. The ozone layer in the stratosphere absorbs UV radiation and creates a warm layer of air in the stratosphere and is therefore responsible for the thermal structure of the stratosphere. Ozone that is present in the troposphere is mostly a result of anthropogenic pollution and therefore higher concentrations are found in urban areas. Ozone is involved with NO_x in the photochemical production of many of the constituents of pollution environments. [also see *stratosphere;* and *troposphere*]

P₂O₅ (phosphorus pentoxide). The designation on a fertilizer label that denotes the percentage of available phosphorus reported as P_2O_5. To convert the percentage of P_2O_5 to percentage of actual phosphorus, multiply the P_2O_5 by 0.437. [Example: 100 pounds of 20-15-10 fertilizer contains: 15 X 0.437 = 6.56 pounds of actual phosphorus] [also see *phosphate;* and *phosphorus*]

Pan. A layer in soils that is strongly compacted, indurated (hardened), or very high in clay content. [also see *cemented (soil); claypan; duripan;* and *hardpan*]

Parasite. An organism that lives on or within a host (another organism); it obtains nutrients from the host without benefiting or killing (although it may damage) the host.

Parasitism. (1) A type of symbiotic relationship in which one organism benefits and the other does not. (2) Feeding by one organism on the cells of a second organism, which is usually larger than the first. The parasite is, to some extent, dependent on the host at whose expense it is maintained.

Parenchyma cells. Thin-walled cells that make up the bulk of most non-woody structures, yet sometimes their cell walls can be lignified. Parenchyma cells in between the epidermis and pericycle in a root or shoot constitute the cortex, and are used for storage of food. They are mainly present in the soft areas of the stems, leaves, root, flowers, fruits etc. Parenchyma cells within the center of the root or shoot constitute the pith. Parenchyma cells in the ovary constitutes the nucellus and their formation is brick-like. Parenchyma cells in the leaf constitute the mesophyll and are responsible for photosynthesis and allow for the interchange of gases. [also see *collenchyma cells*]

Parent material. The unconsolidated and more or less chemically weathered mineral or organic matter from which the surface horizons (layers) of soils are developed by pedogenic processes.

Particle density (soil). The mass per unit volume of the soil particles. Particle density, the density of the solid soil particles only (excluding air), is often expressed as grams per cubic centimeter ($g\ cm^{-3}$) (or $Mg\ m^{-3}$). The standard value used in calculation is 2.65 $g\ cm^{-3}$ since that is the average particle density of the dominant soil minerals, quartz, feldspars, micas and clay minerals. [also see *bulk density, soil*].

Particle-size. The effective diameter of a particle measure by sedimentation, sieving, or micrometric methods. Particle-size distribution is for the whole soil, not just the fine earth fraction. Seven classes are used for most soils.

Seven Particle-Size Classes	
Fragmental	Mostly made up of stones, cobbles, gravel, and very coarse sands without enough fine particles for fill voids larger than 1 millimeter
Sandy-skeletal	More than 35% of material is coarser than 2 millimeters diameter with enough sand to fill voids larger than 1 millimeter
Loamy-skeletal	Same as Sandy-skeletal, except loam fills the voids
Clayey-skeletal	Same as Sandy-skeletal, except clay fills the voids
Sandy	All material is sand or loamy sand, except the very fine sand size fraction
Loamy	All material is between sandy and clayey
Clayey	Material is more than 35% clay. (Fine clayey is 35% to 60% clay; very fine clayey is more than 60% clay

Miller and Gardiner. Soils in Our Environment. 11th ed. 2007.

Parts per million. An expression of concentration. <u>One percent is equivalent to 10,000 ppm</u>. One ppm of nutrient in the soil is equivalent to 2 pounds per acre (assuming ~2,000,000 pounds of soil in the top 6-inch surface layer [acre furrow slice]).

Pasteurization, soil. Soil sterilization kills all organisms in the soil, while pasteurization kills harmful microorganisms and weed seeds. The most common used methods are steam pasteurization, electrical pasteurization. [also see *sterilization, soil*]

Pathogen. A pathogen is an organism, chiefly a microorganism (i.e., viruses, bacteria, fungi, and all forms of animal parasites and protozoa) capable of producing an infection or disease in a susceptible host.

PCB's (polychlorinated biphenyls). A group of synthetic, toxic industrial chemical compounds once used in making paint and electrical transformers. PCBs are chemically inert and not biodegradable. PCBs were frequently found in industrial wastes, and subsequently found their way into surface and ground waters. As a result of their persistence, they tend to accumulate in the environment. In terms of streams and rivers, PCBs are drawn to sediment to which they attach and can remain more or less indefinitely. Although virtually banned in 1979 with the passage of the Toxic Substances Control Act, they continue to appear in the flesh of fish and other animals.

Peat. An accumulation of dead plant material often forming a layer many meters deep. Peat is only slightly decomposed due to being completely waterlogged. Peat is composed chiefly of organic matter that contains some nitrogen of low activity.

Peat soil. (1) An organic soil containing more than 50% organic matter. (2) An organic soil in which the plant residues are recognizable. The sum of the thicknesses of the organic layers are usually greater than the sum of the thicknesses of the mineral layers. [also see *muck soil*]

Pebbles. Rounded or partially rounded rock or mineral fragments between 2 and 75 millimeters in diameter.

Pectin. (1) An important component of the plant cell walls. (2) Any of a group of water-soluble colloidal carbohydrates of high molecular weight found in ripe fruits such as apples, plums, and grapefruit, and used to jell various foods, drugs, and cosmetics.

Ped. A unit of soil structure. An aggregate, such as prism, block, or granule, formed by natural processes. [also see *clod;* and *structure, soil*]

Pedogenesis. The natural process of soil formation.

Pedology. The study of soils as naturally occurring phenomena taking into account their composition, distribution, and method of formation.

Pedon. The smallest unit or volume of soil that contains all the soil horizons of a particular soil type, usually having a surface area of 10.76 square feet or approximately 1 square meter and extending from the ground surface down to bedrock. [also see *profile, soil*]

Pedosphere. The area of the Earth comprised of the solid plates of the continental crust, loose rocks, and soil.

Pedoturbation. The mixing of soil materials or components by natural processes excluding illuviation. [also see *illuviation*]

Pedounit. A selected column of soil containing sufficient material in each horizon for adequate laboratory characterization.

Peneplain. A large flat or gently undulating area. The formation of peneplains are attributed to progressive erosion by rivers and rain, which continues until almost all of the elevated portions of the land surfaces are worn down. When a peneplain is elevated, it may become a plateau which then forms the initial stages in the development of a second peneplain. [also see *plateau*]

Perched water table. Groundwater that is unconfined and separated from an underlying main body of groundwater by an unsaturated zone. Also known as perched groundwater. [also see *water table*]

Periodic Table of the Elements.

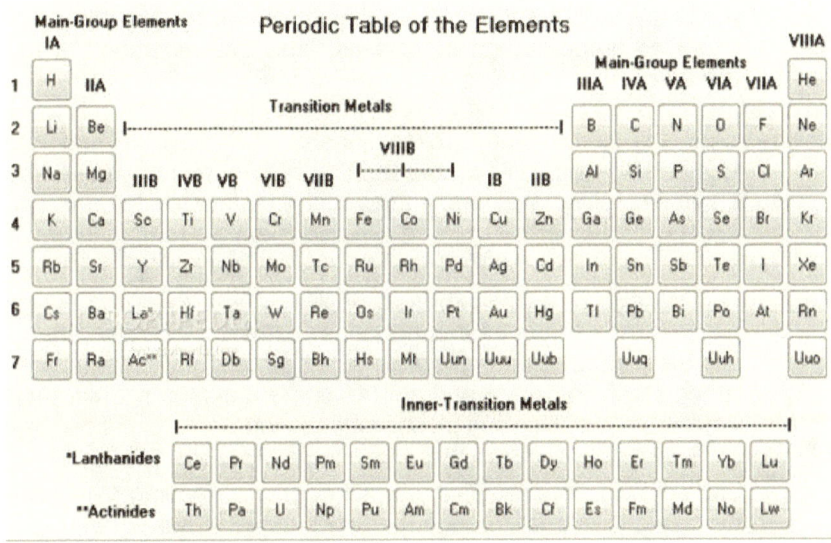

The periodic table of the chemical elements (also periodic table of the elements, or just the periodic table) is a tabular display of the chemical elements. Its invention is generally credited to

Russian chemist Dmitri Mendeleev in 1869, who intended the table to illustrate recurring ("periodic") trends in the properties of the elements. The layout of the table has been refined and extended over time, as new elements have been discovered. The current standard table contains 118 elements to date.

Note: It is frequently quoted that there are "92 naturally occurring chemical elements," but this is incorrect. <u>There are actually only 88 naturally occurring chemical elements</u>. The elements technetium (43), promethium (61), astatine (85) and francium (87) have no stable isotopes, and none of long half-life, so they are not naturally present or occurring. [also see *element;* and *plant nutrients, essential*]

Perennial. A perennial plant or simply perennial (Latin *per*, "through", *annus*, "year") is a plant that lives for more than two years. The term is often used to differentiate a plant from shorter lived annuals and biennials. Woody plants like shrubs and trees are also perennials [also see *biennial;* and *annual*]

Percolation (of soil water). The downward movement of water through soil, especially the downward flow of water in saturated or nearly saturated soil. Percolating water moving through the soil and substrata leaches (the downward movement of free water) dissolved nutrients and other salts. [also see *infiltration, water;* and *leaching*]

Permafrost. (1) Permanently frozen subsoil. (2) Soil, silt and rock located in perpetually cold areas that remain frozen year-round. Though a thin layer may thaw during summer months, the majority of the permafrost in a given location will remain frozen.

Permanent wilting point (percentage) (PWP). The level of moisture in the soil when the plant can no longer extract water from the soil and the plant permanently wilts. The permanent wilting point will change with the plant and soil as some plants can draw more water from a soil, or any given soil, than others.

Soil will hold water against the pull of gravity keeping it available for plants to extract through their root zones. There are limits to the amount of available water. The upper limit is the field capacity (FC) of the soil while the lower limit is the permanent wilting point. <u>Between field capacity and permanent wilting point is the "available" water</u>. This is the water that the plant can pull or draw from the soil through its root system. [also see *field capacity*]

Permeability, soil. The ease with which gases, liquids, or plant roots penetrate or pass through a bulk mass of soil or a layer of soil. Permeability is one of the most important factors determining the overall productivity of a soil or its suitability for development. Permeability is usually limited by the most compacted or clayey layer in the soil profile. The permeability of a soil is controlled by the least permeable horizon even though the others are permeable. Permeability is also significantly affected by amounts of soil organic matter and calcium in relation to sodium plus magnesium in the soil [Ca : (Na + Mg)]. [also see *soil organic matter*, and *calcium*]

USDA Soil Permeability Classes	
Extremely slow	less than 0.01 inches per hour (0.025 cm)
Very slow	0.01 to 0.06 inches per hour (0.025 to 0.15 cm
Slow	0.06 to 0.2 inches per hour (0.15 to 0.51 cm)
Moderately slow	0.2 to 0.6 inches per hour (0.51 to 1.5 cm)
Moderate	0.6 to 2.0 inches per hour (1.5 to 5.1 cm)
Moderately rapid	2.0 to 6.0 inches per hour (5.1 to 15 cm)
Rapid	6.0 to 20.0 inches per hour (15 to 50 cm)
Very rapid	greater than 20 inches per hour (50 cm)

USDA. Keys to Soil Taxonomy. 11th ed. 2010.

Persistence. Refers to a slowly decomposing substance that remains active in the natural cycle for a long period of time.

pH, soil (soil reaction) [from French p (pouvoir) and H (hydrogene); literally: "hydrogen power"]. (1) A symbol denoting the relative concentration of hydronium ions (H_3O^+) [or hydrogen (H^+)] in a solution; pH values run from 0 to 14 with the lower the value the more acidic a solution, i.e., the more hydronium ions it contains.

Soil pH (reaction) Descriptive Terms	
Extremely acid	less than 4.5
Very strongly acid	4.5 to 5
Strongly acid	5.1 to 5.5
Moderately acid	5.6 to 6.0
Slightly acid	6.1 to 6.5
Neutral	6.6 to 7.3
Slightly alkaline	7.4 to 7.8
Moderately alkaline	7.9 to 8.4
Strongly alkaline	8.5 to 9.0
Very strongly alkaline	greater than 9.0

Miller and Gardiner. Soils in Our Environment. 11th ed. 2007.

(2) The degree of acidity or alkalinity (basicity) of a soil, usually expressed as a pH value. Exactly, the negative logarithm of the hydronium ion activity of a soil.

A pH of 7.0 indicates precise neutrality. Values between 7 and 14 indicate increasing alkalinity with pH = 8 being 10 times more alkaline than pH = 7, pH = 9 being 100 times more alkaline than pH = 7, etc. Values between 7 and 0 indicate increasing acidity with pH = 6 being 10 times more acid than pH = 7, pH = 5 being 100 times more acid than pH = 7, etc. Soil pH values of 6.4 are optimum for nutrient availability with most mineral agricultural and horticultural soils in temperate climates (e.g., California); pH values of 5.5 are optimum for organic soils; while optimum pH for tropical soils (e.g., Hawaii) range from 5.5 to 6.5.

Soil pH is used as an initial diagnostic tool in establishing optimum growing conditions for crops and plants. Low soil pH (acid soil) is an indication of a need for liming to diminish acidity and enhance nutrient availability. High soil pH (alkaline soil) is an indication of a need for acidifying soil to correct for harmful levels of sodium, etc., and to also enhance nutrient availability.

Phloem (plural phloems). In a vascular plant, the tissue conducting sugars, and some amino acids, generally downward. Phloem is composed of sieve elements, parenchyma cells, and sometimes fibers and sclereids. [also see *vascular;* and *xylem*]

Phosphate. (1) The amount of pentavalent phosphorus in a fertilizer material calculated as phosphorus pentoxide (P_2O_5) [AAPFCO]. (2) In the fertilizer industry, the term phosphate is usually applied to any phosphate material used as a fertilizer. "Available phosphate" means the sum of the water-soluble and the citrate-soluble phosphate in a fertilizer. In the United States, these are primarily the calcium phosphates and the ammonium phosphates. [also see *monocalcium phosphate; ammonium phosphate; ammonium phosphate nitrate;* and *ammonium phosphate sulfate*]

Phosphate rock. See *apatite*.

Phosphoric acid. (1) Aqueous phosphoric acid (H_3PO_4) widely used in the fertilizer industry. Phosphoric acid is manufactured by treating rock phosphates with sulfuric acid (H_2SO_4) which is then called a green or wet process acid. A higher concentration of H_2SO_4 is used in the wet process acid reaction than is used to produce single superphosphate. However, both reactions produce gypsum ($CaSO_4 \cdot 2H_2O$) which can be used for other industrial purposes and as a soil amendment. (2) The AAPFCO has also adopted as official the following definition: "The term phosphoric acid designates P_2O_5, and refers to the phosphorus content of a fertilizer expressed as P_2O_5 (phosphoric acid, in this case). [also see *superphosphate;* and *gypsum*]

Phosphorus (P). An essential plant nutrient present in soils in organic forms and as iron, manganese, aluminum, and calcium phosphates. <u>Soil pH greatly influences the ratio of different phosphate ions and its total availability.</u> In weakly acidic soil conditions plants take up phosphorus from the soil solution primarily as the dihydrogen phosphate ion ($H_2PO_4^-$); and in weakly basic soil conditions in smaller amounts as the hydrogen phosphate ion (HPO_4^{2-}). However, in strongly basic conditions the phosphate ion (PO_4^{3-}) predominates; and in strongly-acidic conditions aqueous phosphoric acid (H_3PO_4) is the main form.

Typically, only very small quantities of phosphorus are present in the soil solution and these quantities must be continually replenished from other forms of soil phosphorus. Many agricultural and horticultural soils are deficient or low in phosphorus. <u>Since phosphorus is most readily available for plants at soil pH values ≈ 6.4, it is imperative to monitor soil pH to prevent values neither falling below 6.1 nor rising above 7.2 when possible.</u>

Phosphorus is a mobile nutrient within the plant and plays key roles in photosynthesis, respiration (utilization of sugars), energy storage and transfer, cell division, cell enlargement, genetic coding, and many other plant processes. Adequate phosphorus promotes early root formation and growth, improves fruit, vegetable, and grain yield and quality, accelerates maturity, increases resistance to winter kill, improves use efficiency of nutrients such as nitrogen, and increases water use efficiency.

Deficiency symptoms develop early in plant growth. Plants are stunted and may develop dark green, blue, purple or red coloring in older leaves first due to the accumulation of sugars (the purpling/red is an accumulation of anthocyanin pigments) in the plant. Leaves may curl under, turn brown and die. Small-formed buds are another main symptom. Yields are severely diminished. [also see *anthocyanin; P_2O_5 (phosphorus pentoxide); plant nutrients, essential;* and *anion*]

Photic zone. The uppermost layer of a body of water or soil that receives enough sunlight to permit the occurrence of photosynthesis.

Photoautotroph. An organism able to use light as its sole source of energy and carbon dioxide as sole carbon source. Green plants and some bacteria are photoautotrophs. [also see *bacteria, photoautotrophs*]

Photoheterotroph. Organisms that use light for energy, but cannot use carbon dioxide as their sole carbon source. Consequently, they use organic compounds such as carbohydrates, fatty acids, and alcohols as their organic "food" to satisfy their carbon requirements. Examples are purple and green non-sulfur bacteria and heliobacteria.

Photosynthesis. A series of chemical reactions in a plant, using sunlight as energy, that converts carbon dioxide and water into chemical energy...molecules such as glucose that the plant can use as an energy source. The complete balanced equation for photosynthesis is:

$$6CO_2 + 12H_2O \longrightarrow C_6H_{12}O_6 \text{ (glucose)} + 6O_2 + 6H_2O$$

[also see *respiration*]

Phototroph. Organisms that use sunlight to synthesize organic nutrients as their energy source. [Examples: cyanobacteria, algae, and green plants].

Phylloxera. Scientific Name: *Daktulosphaira vitifoliae*. A tiny aphid-like insect that feeds on *Vitis vinifera* grape roots, stunting growth of vines or killing the plants.

Physical (mechanical) weathering (disintegration). One of the two basic weathering processes of rocks and minerals into soil. Disintegration results in a decrease in size of rocks and minerals without appreciably affecting their composition. [also see *chemical weathering*]

Physiological drought. A temporary daytime state of drought in plants due to the losses of water by transpiration being more rapid than uptake by roots although the soil may have an adequate supply. Such plants usually recover during the night.

Phytolith. An opaline formation in plant tissue that remains in the soil after the softer plant tissue has decomposed.

Phytophthora cinnamomi. A destructive parasitic fungi causing root rot in plants.

Phytophthora ramorum. A newly identified plant pathogen that causes Sudden Oak Death in a variety of hosts.

Phytomass. The above ground portion of materials of plant origin usually living, but may also include standing dead trees.

Phytoplankton. Tiny, free-floating, photosynthetic organisms in aquatic systems. Phytoplankton include diatoms, desmids, and dinoflagellates.

Phytotoxic. The property of a substance at a specified concentration that restricts or constrains plant growth.

Phytotoxin. A substance causing growth reduction or death in plants.

Pith. A substance that consists of soft, spongy parenchyma cells, and is located in the center of the stem in eudicots (both herbaceous and woody) and in the center of the roots in monocots. Pith is encircled by a ring of xylem (woody tissue), and outside the xylem, a ring of phloem (bark tissue). In some plants the pith is solid, but for most it is soft. A few plants, such as walnuts, have distinctive chambered pith with numerous short cavities.

Plagioclimax. A plant community which is maintained by continuous human activity of a specific nature, such as burning or grazing.

Plant food. The inorganic compounds within a plant to nourish its cells. A frequent synonym for essential plant nutrients, particularly in the fertilizer trade.

Plant nutrients, essential. Sixteen elements are required by plants to complete their normal life cycles. In addition, nickel, sodium, cobalt and silicon are also sometimes required by certain plant species or associated microorganisms. Their chemical symbols, form(s) common in soil, water, and air and typical composition of plants are listed below. In general, crop use of any nutrient depends on a two-step process: (1) soil supply of that nutrient in an available form, and (2) uptake of the available nutrient by the crop. There are certain constants the grower cannot change. Selecting among the options presented by nature constitutes management. [also see *life cycle*]

Essential Elements Required by Plants			
Element	Chemical Symbol	Portion of Whole Plant (%)	Ion (or Molecule)
Oxygen	O	45-47	CO_2 (mostly through leaves), H_2O, O_2 (mostly through roots)
Carbon	C	41-44	CO_2 (mostly through leaves)
Hydrogen	H	5-6	H_2O (hydrogen from water), H^+
Nitrogen	N	2-4	NH_4^+ (ammonium), NO_3^- (nitrate)

Element	Symbol	Range (%)	Ionic Form
Phosphorus	P	0.3-0.5	$H_2PO_4^-$, HPO_4^{2-} (orthophosphates)
Potassium	K	0.8-1.0	K^+
Calcium	Ca	0.6-2.1	Ca^{2+}
Sulfur	S	0.085-0.4	SO_4^{2-} (sulfate)
Magnesium	Mg	0.3-0.42	Mg^{2+}
Boron	B	0.005 (50 ppm)	H_3BO_3 (boric acid), HB_4O_7, BO_3^-, $B_4O_7^-$
Chlorine	Cl	0.015 (150 ppm)	Cl^- (chloride)
Copper	Cu	0.001 (10 ppm)	Cu^{2+}
Iron	Fe	0.020 (200 ppm)	Fe^{3+} (ferric) [oxidized] Fe^{2+} (ferrous) [reduced]
Manganese	Mn	0.050 (500 ppm)	Mn^{2+}
Molybdenum	Mo	0.0001 (1 ppm)	MoO_4^{2-} (molybdate), $HMoO_4^-$
Zinc	Zn	0.0100 (100 ppm)	Zn^{2+}
Sodium	Na		Na+
Silicon	Si		$Si(OH)_4$ (non-ionized)
Cobalt	Co		Co^{2+}
Nickel	Ni		Ni^{2+}

Compiled From Various Sources

Plastic soil. A moist or wet soil that can be molded without rupture.

Plate. Rigid parts of the Earth's crust and part of the Earth's upper mantle that move and adjoin each other along zones of seismic activity. The theory that the crust and part of the mantle are divided into plates that interact with each other causing seismic and tectonic activity is called plate tectonics.

Plateau. A comparatively flat area of great extent and elevation; specifically an extensive land region considerably elevated [more than 100 meters (approximately 328 feet)] above adjacent lower-lying terrain. Plateaus are commonly limited on at least one side by an abrupt descent and have a flat or nearly level surface. A comparatively large part of a plateau surface is near summit level.

Platy. Soil aggregates that are horizontally elongated. [also see *aggregate*]

Pleistocene period. The geologic period extending from 2,000,000 to 10,000 years before present. In Europe and North America there is evidence of four or five periods of intense cold (or glacial periods) during this time, when large areas of the land surface were covered by ice. During the interglacial periods the climate improved and the glaciers retreated.

Plumule. In botany, the first bud of a young plant; the bud or growing point of the embryo above the cotyledons. [also see *radicle*]

Point source pollution. Water pollution coming from a single point, such as a sewage-outflow pipe. [also see *non-point source pollution*]

Pollutant. A pollutant may be considered as any substance, usually an unwanted by-product or waste, that is released into the environment as a result of (human) activities that alter the chemical, physical and biological characteristics of the environment. These substances may be found in any of the solid, liquid, or gas phases.

Polyacrylamide(s) (PAM). Synthetic water-soluble polymers and copolymers made by polymerizing acrylamide ($CH_2CHCONH_2$) and other monomers. The linear polymers are water-soluble and have application to stabilize structureless, or poorly structured soil, in order to prevent surface crusting, water runoff, erosion, and soil compaction; while improving aeration, friability, pore space, and ease of tillage.

Polymer. A large molecule constructed from many smaller identical units. These include proteins, nucleic acids, and starches.

Polyphosphate(s). Salts or esters of polymeric oxyanions formed from tetrahedral PO_4 (phosphate) structural units linked together by sharing oxygen atoms. When two corners are shared, the polyphosphate may have a linear chain structure or a cyclic ring structure. In biology the polyphosphate esters AMP, ADP, and ATP are involved in energy transfer. Ammonium, calcium and potassium polyphosphates are common examples found in soils. [Example: $(NH_4 PO_3)_n$].

Polysaccharide. A carbohydrate that can be decomposed by hydrolysis (the chemical reaction in which a chemical compound decomposes by reaction with water) into two or more molecules of monosaccharides; especially, any of the more complex carbohydrates (as cellulose, starch, or glycogen). [also see *hydrolysis;* and *monosaccharide*]

Polyvalent cations. [e.g., Ca^{2+}, Mg^{2+}, Fe^{3+}, and Al^{3+}]. These cations play a major role in the stabilization of organic and inorganic colloids. When in abundance, they limit the ability of organic and inorganic colloids to shrink and swell, resulting in a more flocculated (stable) condition. Polyvalent cations serve as bridges between negatively charged clays (inorganic colloids) and negatively charged organic colloids, which aid in structural stability.

Pore space, soil (soil pores).

Pore Size Classification		
Class	Subclass	Class Limits equivalent diameter (μm)
Macropores	Coarse	> 5000
	Medium	2000-5000
	Fine	1000-2000
	Very Fine	75-1000
Mesopores		30-75
Micropores		5-30
Ultramicropores		0.1-5
Cryptopores		<0.1

Brewer, R.. Fabric and Mineral Analysis of Soils. John Wiley & Sons. 1964.

Total space not occupied by soil particles in the bulk volume of soil. The typical ideal mineral soil should contain by volume approximately 50% pore space and 50% soil particles (organic and inorganic). Soil pores have also been referred to as interstices or voids.

Porosity. (1) The volume of the soil mass occupied by pores and pore spaces (not occupied by soil particles). (2) The volume of pores in a soil sample (nonsolid volume) divided by the bulk volume of the sample.

Potable water. Drinkable water, or water suitable for drinking.

Potash (K_2O, potassium oxide). The term potash designates potassium oxide, a strongly corrosive alkali. Soluble potash is that portion of the potassium contained in fertilizer or fertilizer materials that is soluble in aqueous ammonium oxalate, aqueous ammonium citrate, or water, according to an applicable method. (AAPFCO). In the fertilizer industry, potash is used interchangeably with the word potassium and expresses the percentage of potassium oxide (K_2O) in potassium salts and mixtures. K_2O is only 83% actual elemental potassium; therefore, a 100 pound bag of 10-10-10 contains 10% or 10 pounds of nitrogen, 4.4% or 4.4 pounds of elemental phosphorus and 8.3% or 8.3 pounds of elemental potassium. Potash, as used in fertilizer, is usually in the form of sulfate, chloride, nitrate, carbonate or phosphate. [Example: K_2SO_4].

Potassium (K). Potassium is an essential macronutrient and is required by most plants in approximately the same amounts as nitrogen. Approximately 90% of potassium in soils is present in unavailable forms—in primary silicate minerals. Another 2% to 10% is held in slowly available forms between the layers of clay minerals. <u>Available potassium, 1% to 2% of the total, is held on the surface of clay and organic matter colloids and in the soil solution</u>. Many soils are potassium deficient.

Movement of K^+ ions to plant roots is primarily by diffusion through soil water. Therefore, anything that interferes with water movement, such as moisture stress or cold soil temperatures, affects plant uptake of potassium. Plant uptake and plant metabolism of potassium are in the same ionic form (K^+).

Potassium has important roles in activation of many enzyme systems in the plant. It is vital to photosynthesis and to the formation and utilization of sugars (*respiration*). This macronutrient is also essential to protein synthesis and maintenance of protein structure. Potassium helps the plant use water more efficiently and helps control the loss of water from plant leaf surfaces. Adequate potassium helps plants resist diseases and develop strong stems.

Potassium deficiency symptoms, like nitrogen deficiency symptoms, are usually noticeable first in older leaves because of it high mobility. Chlorosis develops around leaf margins. [also see *plant nutrients, essential*; *K_2O (potassium oxide);* and *cation*]

Potassium chloride. See *muriate of potash*.

Potassium nitrate (KNO_3). A fertilizer (also known as "saltpeter") that is chiefly the potassium salt of nitric acid. Potassium nitrate should not contain less than 12% nitrate nitrogen and 44% percent soluble potash (K_2O) [AAPFCO].

Potassium sulfate (sulfate of potash) (K_2SO_4). A fertilizer containing not less than 48% soluble potash (K_2O), chiefly as sulfate, and not more than 2-1/2% percent chlorine.

Potential (or residual) acidity. See *acid soil (soil acidity)*.

ppm. See *parts per million*.

Precipitate. (1) <u>Chemistry</u>: A substance that is separated out from a solution as a solid by the action of chemical reagents, temperature, etc. (2) To cause a slightly soluble substance to become insoluble and separate out from a solution. (3) <u>Meteorology</u>: To condense and fall as rain, snow, sleet, etc.

Precipitation (chemistry). The formation of a solid in a solution or inside another solid during a chemical reaction or by diffusion in a solid. When the reaction occurs in a liquid the solid formed is called the precipitate; or when compacted by a centrifuge, a pellet. The liquid remaining above the solid is in either case called the supernate or supernatant. Powders derived from precipitation have also traditionally been known as flowers.

Precipitation (meteorology). Precipitation often occurs when clouds form upon reaching 100% relative humidity. If condensation nuclei are present, liquid or solid particles of water will form. When this material becomes heavy enough to fall towards the ground it is known as the common phenomenon snow, rain, sleet, or hail. Sometime dew and frost are included as precipitation.

Precision agriculture (prescription farming, site-specific management, precision farming). Placing exact amounts of fertilizers, pesticides, and seeds only where needed for maximum economic yield. The idea with precision agriculture is to know the soil and crop characteristics unique to each part of the field and to optimize the production inputs within small portions of the field. The philosophy behind precision agriculture is that production inputs (seed, fertilizer, chemicals, etc.) should be applied only as needed and where needed for the most economical production.

Primary mineral. (1) A mineral such as feldspar or mica that occurs or occurred originally in an igneous rock. (2) Any mineral that occurs in the parent material of the soil. [also see *secondary mineral*]

Primary plant nutrients (or foods). Often refers to the essential plant nutrients, nitrogen, phosphorus, and potassium; as total nitrogen (N), available phosphate (P_2O_5), and soluble potash (K_2O), which are most often deficient in growing plants unless plants are fertilized. The primary plant nutrients are essential to plant growth in greater quantities than the secondary nutrients and micronutrients. [also see *secondary nutrients; macronutrient(s); plant nutrients, essential;* and *micronutrient(s)*]

Primary producer. A primary producer is any organism that adds biomass to the ecosystem by synthesizing organic molecules from carbon dioxide and simple inorganic nutrients.

Primary root. The first root of the plant, developing in continuation of the root tip or radicle of the embryo. In gymnosperms and dicots the primary root becomes the taproot.

Priming effect. Organic residues with carbon:nitrogen ratios greater than 25:1 often will result in nitrogen deficiencies. This is called "nitrate depression," or is sometimes referred to as the "priming effect." The priming effect will persist until activity of the soil microorganisms decreases due to a lack of carbon. [also see *carbon:nitrogen ratio*]

Producer. Any organism that brings energy into an ecosystem from inorganic sources. Most plants and many protists are producers. [also see *protest*]

Productivity, soil. (1) The output of a specified plant or group of plants under a defined set of management practices. (2) The capacity of a soil to produce a certain yield of crops or other plants with a specified system of management.

Products. The substance(s) that result from a chemical reaction. [Example: In the following chemical reaction, sulfur, water, and oxygen are the reactants that produce the product sulfuric acid]. [also see *reactants*]

$$2S \text{ (sulfur)} + 3H_2O + 3O_2 \longrightarrow H_2SO_4 \text{ (sulfuric acid)}$$

Profile, soil. A vertical section of the soil through all its horizons (or layers) and extending into the parent material. [also see *parent material* and *pedon*]

Prop roots. Adventitious roots (roots originating from the stem, branches, leaves, or old woody roots) arising from the stem above soil level and helping support the plant. [Example: corn (*Zea mays*)]

Protein. Any of a large group of compounds made of linked amino acids containing carbon hydrogen, oxygen, nitrogen, and sulfur, and occasionally some other elements. Proteins are essential in cells of all plants and animals. [also see *amino acid*]

Protista. The taxonomic kingdom from which the other three eukaryotic kingdoms (fungi, animalia, and plantae) are thought to have evolved. The earliest eukaryotes were single-celled organisms that would today be placed in this admittedly not monophyletic group. Protista have a nucleus, large ribosomes, mitochondria, endoplasmic reticulum, and golgi bodies; and many species have chloroplasts. Some protista divide by way of mitosis, meiosis, or both. The majority are single-celled, but nearly every lineage also has multicelled forms. The endosymbiosis theory suggests that eukaryotes may have evolved independently several times. Examples: Amoeba, trypanosoma, plasmodium, giardia and diatoms. [also see *eukaryote*]

Protist. Protists are single-celled eukaryotic organisms [all living organisms except bacteria and blue-green algae (cyanobacteria)] that aren't a plant, animal, or fungus. [also see *eukaryote*]

Protoplasm. (1) The complete cellular contents (cytoplasmic membrane, cytoplasm, and nucleus); usually considered the living portion of the cell, consequently, excluding those layers peripheral to the cell membrane. (2) All the contents of a cell, including the nucleus. [also see *cytoplasm*]

Protozoan (plural protozoa). Unicellular eukaryotic microorganisms (protists) that move by either protoplasmic flow (amoebae), flagella (flagellates), or cilia (ciliates). Most species feed on bacteria, fungi, or detrital particles. Protozoa are abundant wherever bacteria are plentiful. [also see *detritus;* and *eukaryote*]

Puddle. To destroy the structure of the surface soil by physical methods such as the impact of raindrops, poor cultivation practices with implements, and trampling by animals.

Puddled soil. Dense, massive, soil that is artificially compacted when wet; a common result of tillage of clayey soil when wet.

Pupa. In metamorphosing insects, a stage between the larva and adult during which the organism undergoes major developmental changes. [also see *metamorphosis;* and *larva*]

Pyrite (FeS_2). A brass-colored mineral occurring widely and used as an iron ore and in producing sulfur dioxide for sulfuric acid. Also called fool's gold and iron pyrites.

Quality, soil. The capacity of a soil to function within ecosystem boundaries to sustain biological productivity, maintain environmental quality, and promote plant and animal health.

"R" horizon, soil. Underlying consolidated rock with little evidence of weathering.

Radicle. In botany, the radicle is the first part of a seedling (a growing plant embryo) to emerge from the seed during the process of germination. The radicle is the embryonic root of the plant and grows downward in the soil (the shoot emerges from the plumule). [also see *plumule*]

Rainfall interception. The interception and accumulation of rainfall by the foliage and other plant parts of vegetation.

Ratio, fertilizer. See *fertilizer ratio.*

Reactants. The starting substances in a chemical reaction. [Example: In the following chemical reaction, sulfur, water and oxygen are the reactants that produce the product sulfuric acid]. [also see *products*]

$$2S \text{ (sulfur)} + 3H_2O + 3O_2 \longrightarrow H_2SO_4 \text{ (sulfuric acid)}$$

Reaction, soil. See *pH, soil.*

Reactivity, chemical. The state of being chemically reactive. Some atoms react very easily and form compounds so common that we do not find pure deposits of the elements in nature (Example: NaCl, table salt). Other elements are so unreactive that they actually occur in their pure elemental state in nature (Examples: silver and gold). Chemical reactions result in new arrangements of ions and atoms, to form different molecules as shown with table salt:

$$Na + Cl \longrightarrow NaCl \text{ (table salt)}$$

Recalcitrant. (1) Resistant to microbial attack. (2) A term generally applied to organic matter or nutrients that is quite stable and not subject to releasing nutrients into more soluble forms.

Recharge. Water added to an aquifer. [Example: rainfall that seeps into the ground]. [also see *aquifer, water table;* and *perched water table*]

Redox (Reduction-oxidation). A term for the overall reactions in which one substance is oxidized while another is reduced by electron transfers. Redox is usually measured in units of millivolts. [also see *oxidation;* and *reduction*]

> *reduction*
> oxidant + e⁻ ⟶ product
> (electrons gained and oxidation number decreases)
>
> *oxidation*
> reductant ⟶ product + e⁻
> (electrons lost and oxidation number increases)

Reduction. (1) The process by which a compound accepts electrons. (2) Atoms or ions that gain electrons. Reduction is often associated with very wet, waterlogged, soil conditions. [also see *oxidation;* and *redox*]

Example: The reduction of nitrate to dinitrogen in the presence of an acid (denitrification):

$$2\,NO_3^- + 10\,e^- + 12\,H^+ \longrightarrow N_2 + 6\,H_2O$$

Regolith. (1) The layer or mantle of loose, non-cohesive or cohesive rock material that rests on bedrock and forms the surface of the land nearly everywhere. (2) The unconsolidated mantle of weathered rock, soil, and superficial deposits overlying solid rock. Soil scientists only regard the part of the regolith that is modified by organisms and soil-forming processes as soil.

Relative humidity. The ratio between the actual water vapor content of the atmosphere and the maximum water vapor content possible at that given temperature. If the temperature of a given parcel of air rises, the amount of moisture it can hold increases and its relative humidity decreases. If there is no change in temperature but the moisture content decreases, then the relative humidity will again decrease since the ratio of actual water vapor present is less than the maximum amount the air can hold. Conversely, if the water content does not change but the temperature falls, the relative humidity increases until saturation and possible precipitation occurs.

Remediation, soil. Eliminating a contaminant, partially or wholly, to improve a degraded soil.

Residual fertility. See *fertility, residual*.

Residual (or potential acidity). See *acid soil (soil acidity)*

Residual value (benefit). The benefit of fertilizer to succeeding crops after the fertilizer has been in the soil for one or more cropping seasons. Residual values of plant nutrients affect soil test levels for many nutrients (e.g., potassium and phosphorus). Residual values also affect the requirements for any additional nutrients in the upcoming growing year/period. [also see *fertility, residual*]

Residuum (residual soil material). Unconsolidated, weathered, or partly weathered mineral material that accumulates by disintegration of bedrock in place.

Respiration. An intracellular process where molecules, particularly pyruvate in the citric acid cycle, are oxidized with the release of energy. The complete breakdown of sugar or other organic compounds to carbon dioxide and water is termed aerobic respiration, although the first steps of this process are anaerobic. The respiration involving the oxidation of an organic compound (using sugar as an example) is show in the following reaction:

$$C_6H_{12}O_6 \text{ (glucose)} + 6O_2 \longrightarrow 6CO_2 + 6H_2O$$

[Through photosynthesis this reaction is reversed. Carbon dioxide and water are combined by green plants to form sugars; and oxygen (a waste product) is released to be consumed by humans and other animals]. [also see *photosynthesis*]

Retention mechanism. The process by which substances (e.g., nutrients) are retained within the soil profile. [Examples include: precipitation, adsorption, nutrient cycling, and binding into organic matter]

Retrovirus. A virus containing single-stranded RNA as its genetic material and producing a complementary DNA by action of the enzyme reverse transcriptase.

Reversion (of plant nutrients). (1) Changing of essential plant nutrient elements from soluble to less soluble forms by interaction with or reactions in the soil. Reversion is usually restricted to the conversion of monocalcium phosphate ($Ca(H_2PO_4)_2$) to the less soluble dicalcium phosphate ($CaHPO_4 \cdot 2H_2O$). (2) The interaction of a plant nutrient with the soil which causes the nutrient to become less available. [also see *fixation*]

Rhizobacteria. Bacteria that aggressively colonize roots.

***Rhizobia* (singular *Rhizobium*).** Collective common name for bacteria of the genus *Rhizobium* and closely related genera. *Rhizobia* are capable of symbiotic nitrogen fixation, especially with legume plants, from which they receive energy and often fix molecular dinitrogen (N_2) from the atmosphere. [also see *symbiosis*]

Rhizoid. Root-like structures (hairs) that help to hold or anchor a plant to a substrate such as soil, bark, and/or rock.

Rhizome. A more or less horizontal underground stem that produces shoots above and roots below; and is distinguished from a true root by possessing buds, nodes, and usually scale-like leaves. [also see *stolon*]

Rhizomorph. A mass of fungal hyphae organized into long, thick strands usually with a darkly pigmented outer rind and containing specialized tissues for absorption and water transport.

Rhizoplane. (1) Soil in contact with the root surface. (2) Plant root surfaces and usually strongly adhering soil particles. (3) The root surface and its adhering soil.

Rhizosphere. (1) The zone of soil immediately adjacent to plant roots in which the abundance and composition of the microbial population are influenced by the presence of roots. (2) The zone of soil immediately adjacent to plant roots in which the kinds, numbers, or activities of microorganisms differ from that of the bulk soil. The rhizosphere is teeming with bacteria that feed on sloughed off plant cells and the proteins and sugars released by roots. The protozoa and nematodes that graze on bacteria are also concentrated near the rhizosphere.

Rhizosphere competence. The ability of an organism to colonize the rhizosphere.

Rice paddy. An anthropogenic, nearly level impoundment that is inundated with water for long periods; typically for wetland rice production. The term "rice paddy" is applied to areas that have been used in this fashion for a long enough period of time to significantly change the original soil morphology (especially the redoximorphic features).

Rill erosion. See *erosion, rill*.

Riparian. A riparian zone (or riparian area) is the interface between land and a river or stream. Riparian is also the proper nomenclature for one of the fifteen terrestrial biomes of the earth. Plant habitats and communities along the river margins and banks are called riparian vegetation, characterized by hydrophilic plants. Riparian zones are significant in ecology, environmental management, and civil engineering because of their role in soil conservation, their habitat biodiversity, and the influence they have on fauna and aquatic ecosystems, including grassland, woodland, wetland, or even non-vegetative. In some regions the terms riparian woodland, riparian forest, riparian buffer zone, or riparian strip are used to characterize a riparian zone. [also see *hydrophilic*]

Rock(y) outcrop. Surface exposure of bedrock other than lava flows and rock-lined pits.

Root. The usually descending axis of a plant, normally below ground, which serves to anchor the plant and to absorb and conduct water and minerals into it. [also see *adventitious*]

Root cap. The structure that covers and protects the apical meristem in plant roots. Cells forming a protective series of layers over the root meristem.

Root exudates. See *exudates, roots*.

Root hairs. Tubular outgrowths of epidermal cells of the root in the zone of maturation. Root hairs are relatively short-lived and are confined largely to the region of maturation of the root. The production of new root hairs occurs just beyond the region of elongation and at about the same rate as that at which the older root hairs are dying. As the tip of the root penetrates the soil, new root hairs develop immediately behind it providing the root with surfaces capable of absorbing new supplies of water and nutrients from the soil into the plant.

Root interception (of nutrients). The extension (growth) of plant roots into new soil areas where there are untapped supplies of nutrients in the soil and soil solution. [also see *diffusion* and *mass flow*]

Root nodule. A swelling formed on the roots of legume plants, caused by the symbiotic nitrogen-fixing bacteria *Rhizobia*.

Runoff. The portion of precipitation or irrigation on an area that does not infiltrate but instead is discharged from the area. That which is lost without entering the soil is called surface runoff. That which enters the soil before reaching a stream channel is called ground water runoff or seepage flow from ground water. In soil science, runoff usually refers to the water lost by surface flow.

Salination [salinization (obsolete)]. The process whereby soluble salts accumulate in soil. Salination replaces the obsolete term salinization.

Saline soil. A non-sodic soil with a pH value usually less than 8.5, containing sufficient soluble salts to adversely affect the growth of most crop and horticultural plants. The conventional measure of soil salinity is the electrical conductivity of a saturation extract. The lower limit of saturation extract electrical conductivity of such soils is conventionally set at 4 decisiemens per meter (at 25 °C [77 °F]). Salt-sensitive plants are affected at half this salinity and highly tolerant plants at about twice this salinity.

Special treatment including the application of calcium sulfate (anhydrite and/or gypsum) and leaching with good quality irrigation water is generally necessary to remove the excess salts from the root zone. [also see *salinity soil*; *electrical conductivity*; *saline-sodic*; and *sodic*]

Saline-sodic soil. A soil containing sufficient exchangeable sodium to interfere with the growth of most crop plants, plus containing appreciable quantities of soluble salts. The electrical conductivity of the soil solution is greater than 4 decisiemens per meter (dS m^{-1}) (at 25 °C [77 °F]). The exchangeable sodium percentage (ESP) is at least 15, while the sodium adsorption ration (SAR) of the saturation extract is at least 13. The pH of a saline-sodic soil is usually 8.5 or less in the saturated soil.

Special treatment including the application of calcium sulfate and leaching with good quality irrigation water is generally necessary to remove the excess salts from the root zone. [also see *saline soil*; *sodic soil*; and *electrical conductivity*]

Salinity, soil. The quantity of soluble salts in a soil. High levels of salts hinder plant growth and irrigation water infiltration. The conventional measure of soil salinity is the electrical conductivity of a saturation extract; usually expressed in mmhos cm^{-1} (or dS m^{-1}). [also see *electrical conductivity*]

Saline water. Water that contains significant amounts of dissolved solids.

Parameters for Saline Water and Selected Saline Waters of the World (parts per million)	
Fresh water	less than 1,500
Brackish water (slightly saline)	1,500 to 5,000
Saline	from 5,000 to 35,000
Hyper-saline water	from 35,000 to 40,000
Sea/Ocean water	34,527 (Pacific ocean)
Salton Sea, California	44,000
Mono Lake, California	82,000
Great Salt Lake	85,000 (southern portion)
Dead Sea	260,000-350,000
Saltiest Body of Water on Earth Don Juan Pond in Wright Valley, Antarctica, is so salty that it remains liquid even at temperatures as low as -53° C (-63.4° F).	671,000 (42% by weight)

Compiled From Various Sources

Salt(s). (1) Ionic compounds (other than water) containing any ions except OH$^-$ and O^{2-}. Salts are chemical compounds that are usually formed from the combination of an acid and a base in water. Soluble salts found in soils are mostly of the basic cations calcium, sodium, potassium, and magnesium with the acidic anions sulfate, chloride, and carbonate. When mixed with water, a salt may be reactive with other substances. For example, salty water can enhance the corrosion or rusting of steel. Also, some salts can cause burns or irritations on the skin, while others are actually poisonous. Most fertilizers are salts.

Salt index. (1) Salt index is a measure of the relative tendency of a fertilizer to increase the osmotic pressure of the soil solution, as compared to the increase caused by an equal weight of sodium nitrate as a reference material. An excessively high concentration of soluble salts in the soil solution may develop an osmotic pressure exceeding that of the plant sap; and cause dehydration, permanent injury, or even death of the plant. (2) A numerical comparison of fertilizer compounds using sodium nitrate's ($NaNO_3$) Salt Index "100" as the standard.

The higher the salt index the greater the potential for water movement out of the plant issue causing injury or death. Most nitrogen and potash compounds have a high index, and phosphate compounds have a low index.

Examples of Salts			
Salt	Chemical Formula	Derived From	
		Base	Acid
Sodium chloride (table salt)	NaCl ($Na+$ and Cl^-)	NaOH provides Na^+	HCl provides Cl^-
Calcium sulfate (anhydrite)	$CaSO_4$ (Ca^{2+} and SO_4^-)	$Ca(OH)_2$ provides Ca^{2+}	H_2SO_4 provides SO_4^-
Potassium nitrate (fertilizer, etc.)	KNO_3 (K^+ and NO_3^-)	KOH provides K^+	HNO_3^- provides NO_3^-

Salt, table (NaCl). The sodium salt of hydrochloric acid and sodium hydroxide. Pure sodium chloride contains 39% sodium and 61% chlorine.

Salt tolerant. The ability of plants to resist the adverse, nonspecific effects of excessive soluble salts in the rooting medium.

Saltation, erosion. A particular type of momentum-dependent transport especially involving the rolling, bouncing, or jumping action of soil particle 0.1 to 0.5 millimeter in diameter by wind, for relatively short distances and low heights.

Sand. (1) A mineral soil separate consisting of particles greater than 0.05 millimeter and less than 2.0 millimeters in equivalent diameter. (2) A soil textural class. [also see *silt; clay; loam;* and *texture, soil*].

Sandstone. Sedimentary rock composed of sand-sized clasts (A rock fragment or grain resulting from the breakdown of larger rocks).

Sandy. Texture groups consisting of sand and loamy sand textures. [also see *texture, soil*]

Sanitization. The elimination of pathogenic or deleterious organisms, insect larvae, intestinal parasites, and weed seeds.

Sapric soil material. Well decomposed peat. In an "unrubbed" condition, less than 1/3 of the mass is composed of identifiable organic fibers.

Saprophyte. An organism that feeds on dead and decaying organisms allowing the nutrients to be recycled into the ecosystem. Fungi and bacteria are two groups with many important saprophytes.

Saturate. (1) To fill all the voids between soil particles with a liquid. (2) To form the most concentrated solution possible under a given set of physical conditions in the presence of an excess of the solute. (3) To fill to capacity, as the adsorption complex, with a cation species (e.g., Ca^{2+}-saturated, etc.).

Saturated flow. (1) Movement of water through soil by gravity flow, as in irrigation or during a rainstorm. (2) The movement of water in a soil that is completely filled with water.

Saturated soil. A soil where the entire profile is saturated with water.

Saturated soil paste. A particular mixture of soil and water commonly used for measurements (e.g., determining the electrical conductivity of soils) and for obtaining soil extracts. At saturation, the soil paste glistens as it reflects light, flows slightly when the container is tipped and slides freely and cleanly from a spatula for all soils except those with very high clay content.

Saturation percentage (SP). The amount of water required to saturate 100 grams of soil. The value is an estimation of the soil texture and is about twice the field capacity of a soil. [also see *texture, soil;* and *field capacity*] [see table on following page]

Saturation extract. The solution extracted from a soil at its saturation water content.

Saturation Percentages and Estimated Soil Textural Classes	
Saturation Percentage	**Soil Textural Classes**
less than 20	sand or loamy sand
20-30	sandy loam
30-45	loam to silt loam
45-65	clay loam
65-135	clay
135 plus	organic peat or muck

Miller and Gardiner. Soils in Our Environment. 11th ed. 2007.

Scalping. A method of preparing forest soils for planting or seeding that consists of removing the ground vegetation and root mat to expose mineral soil.

Scarp. An escarpment, cliff, or steep slope of some extent along the margin of a plateau, mesa, terrace, or structural bench. A scarp may be of any height. [also see *escarpement*]

Scree. Also called talus, scree is a term given to an accumulation of broken rock fragments at the base of crags, mountain cliffs, or valley shoulders. Landforms associated with these materials are sometimes called scree slopes or talus piles. These deposits typically have a concave upwards form, while the maximum inclination of such deposits corresponds to the angle of repose of the mean debris size. [also see *angle of repose*; *colluvium;* and *talus*]

Screen analysis. A procedure for determining the particle size distribution in a sample of particulate matter, for example, a granular fertilizer. The graduation of test sieve sizes usually conforms to established standard series. The two series most used in the United States are the U.S. Standard Sieve Series, and the Tyler Screen Series.

Sea/Ocean water analysis. Note: If irrigating with ocean/seawater, roughly 547 pounds elemental calcium or 2,378 pounds calcium sulfate (anhydrite or gypsum) [23% calcium content] must be added per 1,000 square feet for every 12 inches irrigation seawater to provide sufficient Ca^{2+} ions to replace excess Na^+ on the CEC sites to replace/leach the sodium. [also see *brackish*]

Sea Water Analysis (Pacific Ocean — California Coast)	
pH	8.0
Electrical Conductivity (ECw)	54 decisiemens per meter (dS m^{-1})
Total Dissolved Solids (TDS)	~34,527 ppm (Pacific Ocean)
Hardness (as CaCO$_3$)	5,494 ppm
Sodium Adsorption Ratio (SAR)	65.76
SARadj	121.66
	parts per million (ppm)
Sodium	11,200
Chlorine	20,078
Bicarbonate	146
Calcium	420
Magnesium	1,304
Potassium	390
Sulfate-sulfur	2,690
Nitrate-nitrogen	12.9
Boron	4.59
Phosphate-phosphorus	<0.1
Iron	0.08
Zinc	<0.01
Manganese	<0.01
Copper	<0.01
Salts applied with seawater irrigation	Each foot of seawater supplies 2,153 pounds of salts per 1000 square feet; or 93,784 pounds (46.9 tons) salts per acre

Compiled From Various Sources

Seaweed. Any large photosynthetic protist, including kelps. Seaweeds are not true plants, but like plants they can make their own food. [also see *kelp;* and *protist*]

Secondary metabolite. A product of intermediary metabolism released from a cell, such as an antibiotic.

Secondary mineral. Minerals that form from the materials released by weathering of other rocks and minerals. The main secondary minerals are the clays and oxides. [also see *primary mineral*]

Secondary plant nutrients. In fertilizers, this refers to calcium, magnesium, and sulfur. The secondary nutrients are essential to plant growth in lesser quantity that the primary nutrients and in greater quantity that the micronutrient elements. [also see *primary plant nutrients; macronutrient(s); micronutrient(s);* and *plant nutrients, essential*]

Sedentary. Living in a fixed location as with most plants, tunicates, sponges, etc.

Sediment. Material, both mineral and organic, that is in suspension, is being transported, or has been moved from its site of origin by water, wind, ice, or mass-wasting and has come to rest on the Earth's surface either above or below sea level. Sediment in a broad sense also includes materials precipitated from solution or emplaced by explosive volcanism as well as organic remains; e.g., peat that has not been subject to appreciable transport. Usually associated with water flow, sediment will accumulate at the mouth of a river or stream as it empties into a larger, slower moving body of water. Sediment deposition is also effected by particle size.

Sedimentary peat. An accumulation of organic material that is predominantly the remains of floating aquatic plants (e.g. algae) and the remains and fecal material of aquatic animals, including coprogeneous earth. [also see *coprogeneous earth;* and *peat*]

Sedimentary rock. A rock formed from materials deposited from suspension or precipitated from solution and usually being more or less consolidated. The principal sedimentary rocks are sandstones, shales, limestones, and conglomerates. [also see *igneous;* and *metamorphic*]

Sedimentation. Separation of a dense material (usually a solid) from a less dense material (usually a liquid) by allowing the denser material to settle out of the mixture.

Seed. A structure produced by a seed-bearing plant [any of the flowering plants (angiosperms) and the conifers and related plants (gymnosperms)] which encapsulates the embryo. The seed often provides nourishment during germination, but may lie dormant for many years first.

Self-mulching soil. A soil in which the surface layer becomes so well aggregated that it does not crust and seal under the impact of rain but instead serves as a surface mulch upon drying. [also see *aggregation, soil*]

Semiarid. Regions or climates where moisture is normally greater than arid conditions but the growth of most crops is still definitely limited. The upper limit of average annual precipitation in the cool semiarid regions is a low as 38 centimeters (15 inches).

Senescence. Advanced age. The complex of aging processes that eventually lead to death. [also see *necrosis*]

Separate, soil.

Soil Separates and Diameter Ranges		
Soil Separate Name	Diameter Range (mm)	Visual Size Comparison of Maximum Size
very coarse sand	2.0 to 1.0	House key thickness
coarse sand	1.0 to 0.5	Small pinhead
medium sand	0.5 to 0.25	Sugar or salt crystals
Fine sand	0.25 to 0.10	Thickness of book page
Very fine sand	0.10 to 0.05	Invisible to the eye
Silt	0.05 to 0.002	Visible under a microscope
Clay	less than 0.002	Most are not visible even with a microscope

Miller and Gardiner. Soils in Our Environment. 11th ed. 2007.

Mineral particles (sand, silt, or clay) less than 2.0 millimeters in equivalent diameter, ranging between specified size limits. [also see *texture, soil*]

Septage. The anaerobic, organic (liquid or solid) waste residues from septic tanks.

Sewage sludge. *See sludge.*

Shale. Sedimentary rock formed by induration of a clay, silty clay, or silty clay loam deposit and having the tendency to split into thin layers.

Sheet erosion. See *erosion, sheet.*

Shoulder. The hillslope position that forms the uppermost inclined surface near the top of a slope. If present, it comprises the transition zone from backslope to the summit of the hill. [also see *catina*; *footslope*; *toeslope;* and *backslope*]

Siemen. A unit of electrical inductance. In SI (international system) metric units electrical conductance is measure in siemens per meter ($S\ m^{-1}$). Decisiemens per meter ($dS\ m^{-1}$) is equivalent to millimhos per centimeter ($mmhos\ cm^{-1}$).

Silica. Amorphous silicon dioxide (SiO_2) (glass). It is a structural component in many organisms such as diatoms, rice plants, and horsetails.

Silicification. The process whereby silica replaces the original material of a substance. [Example: silicified (petrified) wood]

Silicon (Si). An element not found free in nature, but widely distributed as silica (sand or quartz) and silicates in all soils. Silicon actually comprises 27.7% of the weight of the Earth's crust. In comparison, aluminum is only 8.1% of the Earth's crust's mass.

Silt. (a) A mineral soil separate consisting of particles greater than 0.002 millimeter and less than 0.05 millimeter in equivalent diameter. (b) A soil textural class. [also see *sediment*; *sand*; *clay*; *loam;* and *texture, soil*]

Silting. The deposition of water-borne sediments in stream channels, lakes, reservoirs, or on floodplains; usually resulting from a decrease in the velocity of the water. The word/term is often used incorrectly and synonymously with sediment or sedimentation. Silting often includes particles from clay to sand size. [also see *sediment*]

Silviculture. The practice of controlling the establishment, growth, composition, health, and quality of forests to meet diverse needs and values of the many landowners, societies, and cultures. The name comes from the Latin *silvi-* (forest) + culture (as in growing).

Site. (1) In ecology, an area described/defined by its biotic, climatic, and soil conditions as related to its capacity to produce vegetation. (2) An area sufficiently uniform in biotic, climatic, and soil conditions to produce a particular climax vegetation.

Slag, agricultural. A fused silicate whose calcium and/or magnesium content is capable of neutralizing soil acidity and which is sufficiently fine enough to react readily in the soil.

Slick spots. (1) Areas having a puddle or crusted, very smooth, nearly impervious surface. The underlying material is dense and massive. (2) Small areas of surface soil that are slick when wet because of alkalinity or high exchangeable sodium.

Slickenside. The polished surface that forms when two peds rub against each other when some soils expand in response to wetting and drying.

Slime molds. Any of various organisms of the phylum Myxomycota that grow on decaying vegetation and in moist soil and have a similar but more advanced life cycle. Also called myxomycete

Sludge. A general term for solid wastes usually collected by sedimentation from water. Common sludges are sewage, food-processing, and sugar-processing sludge. Primary sludge is produced by sedimentation processes, while secondary sludge is the product of microbial digestion.

Slow-release (fertilizer). *See fertilizer, slow-release.*

Slurry. A thin watery mixture of a fine, insoluble material.

Slurry (fertilizer). *See fertilizer, slurry.*

Sodic soil. A non-saline soil containing sufficient exchangeable sodium to adversely affect crop production and soil structure. The exchangeable sodium percentage (ESP) is at least 15 (exchangeable sodium exceeds 15% of the cation exchange capacity), while the sodium adsorption ration (SAR) of the saturation extract is at least 13. [Note: with sodic soils the pH is usually high, often above 9.0, and plant nutritional imbalances may occur. (A soil pH above 8.4 typically indicates that a sodium problem exists.)]

Saline, sodic, and saline-sodic soils are all terms used to define subjective limits of salt content and exchangeable sodium percentage in salt-affected soils. Sodic soils, with low soluble salts but high exchangeable sodium, tend to remain in a dispersed condition and are almost impermeable to both irrigation and rainwater. Sodic soils are plastic and/or sticky when wet and form hard crusts when drying. When wet, sodic soils have a characteristic smooth, slick look caused by the dispersed condition of clay and humus. Sodic soils are very poor for the growth of most plants.

Special treatment including the application of calcium sulfate (anhydrite and/or gypsum) and leaching with good quality irrigation water is necessary to remove the excess sodium from the root zone. [also see *alkaline soil*; *exchangeable sodium percentage*; *saline soil;* and *saline-sodic soil*]

Sodication. The process whereby the exchangeable sodium content of a soil is increased.

Sodium (Na). A metallic element (Na from Latin *natrium*) and atomic number 11, sodium comprises about 2.8% of the Earth's surface. A soft, silvery-white, highly reactive metal, sodium is a member of the alkali metals within "group 1". Elemental sodium does not occur naturally on Earth because it quickly oxidizes in air and is violently reactive with water. Sodium ions are soluble in water and is therefore present in great quantities in the Earth's oceans and other stagnant bodies of water. In these bodies it is mostly counterbalanced by the chloride ion, causing evaporated ocean water solids to consist mostly of sodium chloride, or common table salt (NaCl). Sodium ions are also a component of many minerals. Sodium is an essential element for all animal life (including humans) and for some plant species.

Sodium adsorption ratio (SAR).

$$SAR = \frac{[sodium]}{[calcium + magnesium]^{1/2}}$$

(1) The tendency for sodium cations to be adsorbed at cation-exchange sites in soil at the expense of other cations, calculated as the ratio of sodium to calcium and magnesium in the soil (as the amount of sodium divided by the square root of half the sum of the amounts of calcium and magnesium, where ion concentrations are given in milliequivalents per liter [meq/L]). A low sodium content gives a low SAR value. In practice, allowance must be made for other reactions within the soil which do not involve sodium but affect concentrations of calcium and magnesium. The SAR value is most likely to be changed by irrigation water. (2) A value representing the relative hazard of irrigation water because of a high sodium content relative to its calcium plus magnesium content. [also see *exchangeable sodium percentage*]

Sodium Hazard of Soil Based on SAR Values				
Classification	Sodium adsorption ratio (SAR)[2]	Electrical conductivity (dS m^{-1})[1]	Soil pH	Soil physical condition
Sodic	>13	<4.0	>8.5	poor
Saline-Sodic	>13	>4.0	<8.5	normal
High pH	<13	<4.0	>7.8	varies
Saline	<13	>4.0	<8.5	normal

[1] dS m^{-1} = mmho cm^{-1}

[2] If reported as exchangeable sodium percentage or ESP, use 15% as threshold value.

Brady and Weil. Elements of the Nature and Properties of Soils. 3rd ed. 2008.

Sodium nitrate (NaNO$_3$). A fertilizer that is chiefly the sodium salt of nitric acid. It should contain not less than 16% nitrate-nitrogen and 26% sodium. Sodium nitrate is seldom used since the element sodium is detrimental to favorable soil structure.

Soil. (1) The unconsolidated mineral (weathered rocks and minerals) and/or organic material on the immediate surface of the Earth that serves as a natural medium for the growth of land plants. (2) Unconsolidated mineral or organic matter on the surface of the Earth that has been subjected to and influenced by genetic and environmental factors of: parent material, climate (including water and temperature effects), macroorganisms and microorganisms, and topography; all acting over a period of time and producing a product--soil--that differs from the material from which it is derived in many physical, chemical, biological, and morphological properties, and characteristics.

Soil, acid (soil acidity). *See acid soil (soil acidity).*

Soil atmosphere. See *atmosphere, soil.*

Soil biochemistry. See *biochemistry, soil.*

Soil compaction. See *compaction, soil.*

Soil horizon. See *horizon, soil.*

Soil organic matter. See *organic matter, soil.*

Soil porosity. See *porosity, soil.*

Soil quality. See *quality, soil.*

Soil Science. The science dealing with soils as a natural resource on the surface of the Earth including soil formation, classification and mapping, and physical, chemical, biological, and fertility properties of soils per se; and those properties in relation to their use and management.

Soil series. Established by the National Cooperative Soil Survey of the United States Department of Agriculture (USDA) Natural Resources Conservation Service, soil series are a level of classification in the USDA Soil Taxonomy classification system hierarchy. The actual object of classification is the so-called soil individual, or pedon. Soil series consist of pedons that are grouped together because of their similar pedogenesis, soil chemistry, and physical properties. More specifically, each series consists of pedons having soil horizons that are similar in soil color, soil texture, soil structure, soil pH, consistence, mineral and chemical composition, and arrangement in the soil profile. These result in soils which perform similarly for land use purposes.

Soil solution. See *solution, soil.*

Soil structure. See *structure, soil*.

Soil texture. See *texture, soil*.

Soil water potential. *See water potential, soil.*

Solid waste. See *waste, solid*.

Solar radiation. Solar radiation is electromagnetic radiation (light energy) emitted by the sun. This energy is transmitted through space in the units of electromagnetic energy called photons. The strength of the solar energy that reaches our outer atmosphere is called the solar constant and has a value of approximately 2.0 calories per minute per square centimeter.

Solarization. The method to control pathogens and weeds where moistened soil in hot climates is covered with transparent polyethylene plastic sheets, in so doing trapping incoming radiation.

Solubility. The amount of a substance that dissolves in a given quantity of solvent (such as water) at a given temperature to give a saturated solution. [Example: gypsum ($CaSO_4 \cdot 2H_2O$) comprises a solubility of 0.205 grams per 100 grams water]

To be available to plants a nutrient must be at least slightly soluble in the soil solution. Solubility of compounds such as urea, ammonium nitrate, and potassium chloride increases rapidly with temperature. The presence of other substances in the solution may either increase or decrease the solubility. [also see *fertilizer, solubility of some fertilizers*]

Soluble. Capable of being dissolved in a solvent (usually water).

Soluble potash. See *water soluble potash (K_2O)*.

Soluble salts. (1) An ionic compound that dissolves in a solvent (usually water). (2) Salts that can be removed from the soil by water alone. [Example: sodium chloride (NaCl)]. [also see *exchangeable ion*]

Solum (plural sola). The upper and most weathered part of the soil profile. The A, E, and B horizons.

Solute. In the case of a solution of a solid dissolved in a liquid, the solid. [also see *solution*; and *solvent*]

Solution. (1) A homogeneous (uniform) molecular mixture, usually a liquid. (2) The act of process of dispersing one or more substances in another, usually a liquid, so as to form a homogenous mixture; a dissolving.

The substance present in the greatest amount, usually a liquid, is called the **solvent**. The substances present in lesser amount are called **solutes**. Many substances within living systems are found in solution.

Solution, soil. The aqueous liquid phase of the soil and its solutes. [also see *solute*]

Solvent. In a solution of a solid in a liquid, the liquid. [also see *solute*]

Sorption. (1) A class of processes by which one material is taken up by another. (2) The removal of an ion or molecule from solution by adsorption and absorption. Absorption refers to the process of the penetration of one material into another. Adsorption refers to the action of one material being collected on another's surface. Sorption is often used when the exact nature of the mechanism of removal is not known. [also see *adsorption*; *absorption*; and *desorption*]

Spatial variability. Variation in soil properties (1) laterally across the landscape, at a given depth, or with a given horizon, or (2) vertically downward through the soil.

Sphagnum. The peat moss (division *Bryophyta*) forms extensive bogs in cold and temperate regions of the world.

Spawn. The term used for the combination of substrate plus mycelium in mushroom farming. The spawn can be thought of as the vegetative part of the mushroom.

Species. In microbiology, a collection of closely related strains sufficiently different from all other strains to be recognized as a distinct unit.

Spent mushroom substrate (SMS) [compost]. See *mushroom compost*.

Spent lime (sugar beet lime) (CaCO$_3$). Spent lime is a by-product of the sugar beet purification process used as an agricultural liming product. It is generated by heating mined calcium carbonate limestone to form calcium oxide and carbon dioxide. These two products are injected into the thick beet juice during processing where they reform as calcium carbonate (CaCO$_3$). When the calcium carbonate reforms it captures and adsorbs many of the impurities in the juice and precipitates out from the juice. The precipitate forms a solid lime product that needs to be discarded, leaving behind the thin juice from which sugar is extracted. The calcium carbonate equivalent for spent/sugar beet lime is 80-90%. [also see *lime, agricultural*]

Splash erosion. See *erosion, splash*.

Spore. (1) Specialized reproductive cell. Asexual spores germinate without uniting with other cells, whereas sexual spores of opposite mating types unite to form a zygote before germination occurs. (2) Usually a single cell that is dispersed as a reproductive unit, cell, or body produced in plants, bacteria, and protozoa. Spores are capable of developing into an adult without fusion with another cell.

Springtails. See *Collembola*.

Starch. A complex, high-energy, polymer of glucose used by plants and green algae to store surplus sugar for later use. The basic chemical formula of the starch molecule is $(C_6H_{10}O_5)_n$.

Stele. In a vascular plant, the stele is the central part of the root or stem containing the tissues derived from the procambium. These include vascular tissue, in some cases ground tissue (pith) and a pericycle, which, if present, defines the outermost boundary of the stele. The endodermis is the innermost cell layer of the cortex. [also see *casparian strip*]

Sterilization, soil. The act of destroying all living organisms in the soil. Rendering a soil free of viable micro- and macroorganisms. [also see *pasteurization, soil*]

Stomate (or stomata) [plural stoma]. A minute opening bordered by guard cells in the epidermis of leaves and stems through which gases (e.g., especially carbon dioxide and oxygen) pass. Also used to refer to the entire stomatal apparatus; the guard cells plus their included pore. [also see *guard cells*]

Stratosphere. The thermal atmospheric region of the atmosphere between the troposphere and the mesosphere. The lower boundary of the stratospheric region is marked by the tropopause and begins at approximately 13 kilometers; however, this altitude of the troposphere depends on latitude. The upper limit of the stratosphere is marked by the stratopause at approximately 50 kilometers. The stratosphere is characterized by relatively stable temperatures in the lower regions (between -80 °C and -50 °C), and begins warming near 20 kilometers, reaching its maximum temperature of approximately 0 degree Celsius at the stratopause. [also see *atmosphere*; *troposphere;* and *mesosphere*]

Stratum (plural strata). A layer of sedimentary rock.

Stream terrace. One of a series of platforms in a stream valley, flanking (and more or less parallel to) the stream channel; originally formed near the level of the stream and representing the dissected remnants of an abandoned flood plain, stream bed, or valley floor produced during a former state of erosion or deposition.

Strip cropping. The practice of growing crops for controlling erosion that requires different types of tillage in alternate strips along contours or at right angles to the prevailing direction of erosive winds.

Stolon. A stem that grows horizontally along the ground surface and may form adventitious roots, such as the runners of a strawberry plant. A method of plant vegetative propagation. [also see *rhizome;* and *adventitious*]

Stoma. See *stomate*.

Structure, soil. The combination or arrangement of primary soil particles into secondary particles, units, or peds. The secondary units are characterized on the basis of size, shape, and grade (degree of distinctness). [also see *structure; soil grades; texture, soil;* and *ped*]

Structure, soil grades. A grouping or classification of soil structure on the basis of inter- and intra-aggregate adhesion, cohesion, or stability. Four grades of structure are recognized as follows:

Structureless: No observable aggregation or no definite and orderly arrangement of natural lines of weakness. Massive, if coherent; single-grain, if noncoherent.

Weak: Poorly formed indistinct peds barely observable in place. When gently disturbed, the soil material parts into a mixture of whole and broken units and much material that exhibits no planes of weakness.

Moderate: Well-formed distinct peds evident in undisturbed soil. When disturbed, soil material parts into a mixture of whole units, broken units, and material that is not in units.

Strong: Peds are distinct in undisturbed soil. They separate cleanly when soil is disturbed, and the soil material separates mainly into whole units when removed.

Stubble mulch. *See mulch, stubble.*

Subaerial. Located or occurring on or near the surface of the Earth.

Subsoil. (1) B horizons are commonly referred to as the subsoil. They are a zone of accumulation where rain water percolating through the soil has leached material from above and it has precipitated within the B horizons, or the material may have weathered in place. (2) That part of the soil below the plow layer. (3) The underlying layers of the soil beneath the topsoil that may contain less soil organic matter and are generally characteristic of the soil's parent material. Subsoils tend to have lower available nutrients, especially nutrients such as phosphorus and the metal micronutrients. [also see *"B" horizon, soil*]

Subsoiling. The tillage of subsurface soil, without inversion, to break up dense soil layers that restrict water movement and root penetration. Subsoiling is typically achieved with the use of chisel and/or ripper tillage equipment.

Substance, chemical. In chemistry, a material with a specific chemical composition. A common example of a chemical substance is pure water; it has the same properties and the same ratio of hydrogen to oxygen whether it is isolated from a river or made in a laboratory. Some typical chemical substances are diamond, gold, salt (NaCl) and sugar (sucrose, $C_{12}H_{22}O_{11}$). Generally, chemical substances exist as a solid, liquid, gas, or plasma and may change between these phases of matter with changes in temperature or pressure. Chemical reactions convert one chemical substance into another.

Substrate. (1) That which is laid or spread under an underlying layer, such as the subsoil. (2) The substance, base, or nutrient on which an organism grows. (3) Compounds or substances that are acted upon by enzymes or catalysts and changed to other compounds in a chemical reaction.

Succession. A gradual process brought about by the change in the number of individuals of each species of a community and by the establishment of new species that gradually replace the original inhabitants.

Sucrose. Common table sugar. A sweet white crystalline solid often used as a food additive. The chemical formula for sucrose is $C_{12}H_{22}O_{11}$ and its molar mass is 342.30 g mol^{-1}. The solubility of sucrose in water at 20 °C is 211.5 grams per 100 ml.

Sugar beet lime. See *spent lime*.

Sulfur (S). Sulfur is an essential secondary plant nutrient classed with calcium and magnesium. Sulfur exists in the soil in a number of oxidation states and is absorbed by the plant in the sulfate ion (SO_4^{2-}) form. Sulfate is reduced in the plant before incorporation into plant components. The oxidation of elemental sulfur (S) is show below:

$$2S + 2H_2O + 3O_2 \xrightarrow{\textit{Thiobacillus}} 4H^+ + 2SO_4^{2-} \text{ (sulfate)}$$

Sulfur is essential in forming plant protein because it is part of several amino acids (cystine, methionine, and cysteine). As a part of plant protein, it is essential for enzyme activity. Sulfur is also involved in nodule formation and nitrogen fixation in legumes. Sulfur is essential in chlorophyll formation although it is not a constituent of the chlorophyll molecule. Sulfur deficient plants are pale green. Symptoms look similar to nitrogen deficiency, however, since sulfur in not mobile like nitrogen, symptoms generally appear first on the upper, new leaves, while nitrogen deficiency shows up first on the lower, older leaves. In sulfur deficiency, the entire plant can take on a pale green appearance. Sulfur deficiencies occur most often in sandy soils low in organic matter. Crop removal varies from about 10 pounds per acre for grain crops to about 20 pounds per acre for legumes.

As a soil amendment, elemental sulfur (S) is widely used to acidify high pH soils as shown in the following reactions:

$$2S + 3O_2 + 2H_2O \longrightarrow 2H_2SO_4 \text{ (sulfuric acid)}$$

$$H_2SO_4 + CaCO_3 \text{ (lime)} \longrightarrow CaSO_4 + CO_2 \text{ (gas)} + H_2O$$

[also see *plant nutrients, essential*; *anion*; and *amendment, soil*]

Approximate Amounts of Soil Sulfur (99%) Required (To Increase Soil Acidity of Top Six Inches of Carbonate-Free Soil)			
Desired pH change	Sulfur (S) Required per Acre (pounds)		
	Sand	Loam	Clay
8.5 to 6.5	2,000	2,500	3,000
8.0 to 6.5	1,200	1,500	2,000
7.5 to 6.5	500	800	1,000
7.0 to 6.5	100	150	300

Thorup. Ortho Agronomy Handbook. 1984.

Sulfur cycle. The sequence of transformations undergone by sulfur where it is taken up by living organisms, transformed upon death and decomposition of the organism, and converted ultimately to its original state of oxidation.

Summer fallow. *See fallow.*

Superphosphate [$Ca(H_2PO_4)_2 \cdot H_2O$]. A fertilizer product obtained when phosphate rock is treated with sulfuric acid (H_2SO_4), phosphoric acid (H_3PO_4), or a mixture of these acids. The guaranteed percentage of available phosphate is to be stated as a part of the name. [also see *phosphoric acid; monocalcium phosphate, ($Ca(H_2PO_4)_2$; and triple superphosphate*]

Normal. Also called ordinary or single superphosphate. Superphosphate made by reaction of phosphate rock with sulfuric acid, usually containing 7% to 10% P (16% to 22% P_2O_5).

Concentrated. Also called triple or treble superphosphate, made with phosphoric acid and usually containing 10% to 21% P (44% to 48% P_2O_5).

Surface area. The area of the solid particles in a given quantity of soil or porous medium.

Surface sealing, soil. The deposition by water, orientation and/or packing of a thin layer of fine soil particles on the immediate surface of the soil, greatly reducing its water permeability. Soil surface sealing can be ameliorated with the addition of soil amendments such as composts and other organic materials, and with the use of calcium sulfate in the form of anhydrite and/or gypsum. [also see *anhydrite*; *gypsum;* and *permeability*]

Surface soil. The uppermost part of the soil ordinarily moved in tillage, or its equivalent in uncultivated soils and ranging in depth from 7 to 25 centimeters (approximately 2.75 to 10 inches). Frequently designated at the plow layer, the surface layer, the Ap layer, or the Ap horizon. [also see *topsoil*]

Surfactant. From: **SURF**ace **ACT**ive **A**ge**NT**. (1) A substance that exists at the boundary between two other substances. For example, detergents have one end highly soluble in greasy, non-polar substances and one end soluble in water. (2) A substance that lowers the surface tension of a liquid. There are three types of surfactants: (1) nonionic, (2) anionic, and (3) cationic.

Suspension fertilizers (liquid). See *fertilizers, suspension*.

Sustainability. Managing soil and crop cultural practices so as not to degrade or impair environmental quality on or off site; and without eventually reducing yield potential as a result of the chosen practice through exhaustion or either onsite resources or non-renewable inputs.

Sustainable agriculture. An integrated system of plant and animal production practices having a site-specific application that will, over the long term:
- satisfy human food and fiber needs

- enhance environmental quality and the natural resource base upon which the agricultural economy depends
- make the most efficient use of nonrenewable resources and on-farm resources and integrate, where appropriate, natural biological cycles and controls
- sustain the economic viability of farm operations
- enhance the quality of life for farmers and society as a whole

Sustainable agriculture <u>does not</u> mean a return to either the low yields or poor farmers that characterized the 19th century. Rather, sustainability builds on current agricultural achievements, adopting a sophisticated approach that can maintain high yields and farm profits without undermining the resources on which agriculture depends. In general, sustainable agriculture is the integration of soil and crop management technologies to produce quality food and fiber while maintaining or improving soil productivity and environmental quality. [also see *environmental sustainability*]

Symbiont. An organism in a symbiotic relationship; also called a symbiote. [also see *facultative symbiont;* and *obligate symbiont*]

Symbiosis. Two dissimilar organisms living together in intimate association resulting in mutual benefit, such as *Rhizobia* species and legume plants, and the association of algae and fungi in lichens. [also see *mutualism*]

In symbiosis, the symbiont organisms rely upon one another and both benefit by the relationship. Examples: (1) In the symbiotic relationship of legumes and *Rhizobia* species, the host plants provide fixed carbon sources for the bacteria, which in turn fix atmospheric nitrogen for the host plant. (2) The mycorrhizal, fungus-plant root association. Plant uptake of such nutrients as phosphorus and zinc can be benefited by this type of relationship. [also see *Rhizobia*; *lichens;* and *synergism*]

Symbiotic bacteria. In plant science, the definition usually relates to bacteria in nodules growing on the roots of legumes (especially *Rhizobia*) that have the ability to "fix" molecular nitrogen (N_2) from the atmosphere (the principal component of air) into forms that can be utilized by the host legume plant. This is true symbiosis as neither plant nor bacterium utilizes atmospheric nitrogen in the absence of the second organism.

Unlike oxygen, nitrogen is rather unreactive as a gas due to its relative stability. This stability generally prevents plants from utilizing nitrogen directly from air into their various metabolic processes. [also see *legume;* and *nodule*].

Synergism. (1) The nonobligatory association between organisms that is mutually beneficial. Both populations can survive in their natural environment on their own although, when formed, the association offers mutual advantage. (2) The simultaneous actions of two or more factors that have a greater total effect together than the sum of their individual effects. [also see *symbiosis*]

Synthetic. A term which that be applied to any substance generated from another material or materials by means of a chemical reaction.

Synthetic materials. Materials that are manufactured chemically (by synthesis) from their elements or other chemicals, as opposed to those found in nature.

Syntrophy. The interaction of two or more populations that supply each other's nutritional needs.

Talus. Rock fragments of any size or shape (usually coarse and angular) derived from and lying at the base of a cliff or very steep rock slope. The accumulated mass of such loose, broken, rock formed chiefly by falling, rolling, or sliding. [also see *angle of repose*; *colluvium;* and *scree*]

Tankage. The rendered, dried, and ground by-product, largely meat and bone, from animals slaughtered or that have otherwise died.

Fish tankage. Fish scrap, dry ground fish, fish meal (fertilizer grade) that is the dried ground product derived from rendered or unrendered fish. [also see *fish emulsion*]

Acidulated fish tankage (fish scrap). The rendered product derived from fish and treated with sulfuric acid.

Taproot. The primary root of a plant formed in direct continuation with the root tip or radicle of the embryo. The taproot forms a stout, tapering, main root from which smaller, lateral branches develop.

Temperate. A region in which the climate undergoes seasonal change in temperature and moisture. Temperate regions of the Earth lie primarily between 30 and 60 degrees latitude in both hemispheres.

Tensiometer. A device for measuring the soil water matric potential *in situ*. A porous, permeable, ceramic cup connected through a water-filled tube to a manometer, vacuum gauge, pressure transducer, or other pressure measuring device.

Terrace. (1) A natural level plain bordering a river, lake, or sea. (2) A raised, level, strip of earth usually constructed on or nearly on a contour designed to make the land suitable for tillage and to prevent accelerated erosion.

Terrestrial. Living on land, as opposed to marine or aquatic.

Terric layer. An unconsolidated mineral substratum not underlain by organic matter, or one continuous unconsolidated mineral layer (with 17% or less organic carbon), more than 30 centimeters thick in the middle or bottom tiers; underlain with organic matter and within a depth of 1.6 meters from the surface.

Texture, fine. Soil containing equal to or greater than 35% clay.

Texture, soil. Relative proportions of the various soil size separates.

U.S. Department of Agriculture Soil Texture Classification System	
Very coarse sand	2.0 - 1.0 mm
Coarse sand	1.0 0.5 mm
Medium sand	0.5 - 0.25 mm
Fine sand	0.25 - 0.10 mm
Very fine sand	0.10 - 0.05 mm
Silt	0.05 - 0.002 mm
Clay	<0.002 mm

USDA. Keys to Soil Taxonomy. 11th ed. 2010.

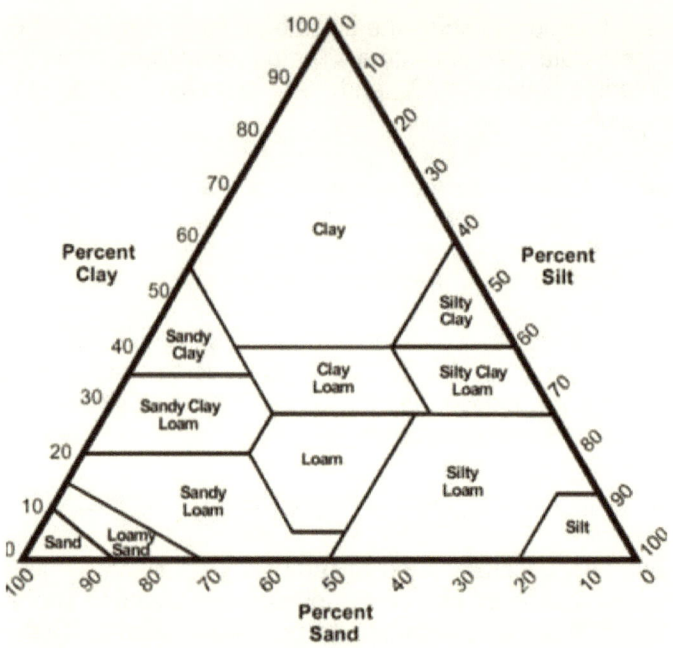

Example: Soil Separates and Textural Classes [using soil textural triangle above]			
% Clay	% Silt	% Sand	Textural Class
20	40	40	Loam
8	10	82	Loamy sand
35	52	13	Silty clay loam

Thallus. A vegetative body that is not differentiated into tissue systems or organs.

Thatch. A mat of non-decomposed plant material (e.g., grass roots) accumulated next to the soil in a grassy area (e.g., a lawn).

Thermophile. An organism whose optimum temperature for growth is between 45 °C and 85 °C (113 °F to 185 °F).

Thermophilic. Heat loving. Especially applied to certain bacteria requiring high temperatures for normal development.

Thermosphere. The outermost shell of the atmosphere, between the mesosphere and outer space, where temperatures increase steadily with altitude. [also see *atmosphere*; *mesosphere;* and *stratosphere*]

Thixotropic (thixotropy). (1) A property of certain gels/fluids that liquefy when subjected to vibratory forces like simple shaking, and then solidify again when allowed to stand undisturbed. (2) The tendency for the viscosity of a liquid to decrease when subjected to shear. Mayonnaise is a good example of a thixotropic fluid. At rest, a thixotropic substance will not pour but will flow readily when submitted to shear. Some silty soils have thixotropic properties.

Tile drain. Lines of concrete, ceramic (clay), plastic pipe, or related structure placed in the subsoil to enhance and/or accelerate drainage of water from the soil profile to an outlet.

Tillage. The mechanical manipulation of the soil profile for any purpose. In agriculture, it is usually restricted to modifying soil conditions and/or managing crop residues and/or weeds and/or incorporating chemicals for crop production.

Tilth. The physical condition of soil as related to its ease of tillage, fitness as a seedbed, and impedance to seedling emergence and root penetration. Tilth generally is a result of tillage.

Toeslope. The hillslope position that forms a gently inclined surface at the base of a slope. [also see *catina*; *shoulder*, *footslope;* and *backslope*]

Topsoil. (1) The "A" horizon. (2) The layer of soil moved in cultivation. (3) Commonly, fertile soil material used to top-dress gardens and lawns. [also see *"A" horizon*]

Tortuosity. The non-straight nature of soil pores.

Total Dissolved Solids (TDS), irrigation water. Solids in water that can pass through a filter (usually with a pore size of 0.45 micrometers). TDS are comprised of inorganic minerals, salts, metals, cations, or anions (predominantly calcium, magnesium, potassium, sodium, carbonates, bicarbonates, chlorides and sulfates) and some small amounts of organic acids that are dissolved in the water.

In general, the TDS concentration is the sum of the cations (positively charged) and anions (negatively charged) ions in the water. The TDS test provides a qualitative measure of the amount of dissolved ions (salts are chemical ionic compounds), but does not tell us the nature or ion relationships. Therefore, the <u>TDS test is used as an indicator test to determine the general quality of irrigation water</u>.

TDS is usually reported as parts per million. Evaporating a known weight of water sample and weighing the salt/organic matter residue remaining can determine this. However, TDS is often approximated by measuring the electrical conductivity (ECw) of the water in decisiemens per meter (dS m^{-1}) and multiplying by 640. This gives the approximate TDS in ppm.

Example: An irrigation water sample has an ECw of 2.50:

2.50 X 640 = 1,600 ppm total dissolved solids (TDS)

TDS is also used to estimate the quality of drinking water since it represents the amount of ions in the water. Drinking water with high TDS often has a bad taste and/or high water hardness, and can result in a laxative effect. The U.S. Environmental Protection Agency has set a standard of 500 mg L^{-1} (ppm) TDS for drinking water.

To measure TDS, the water sample is filtered, and then the filtrate (the water that passes through the filter) is evaporated in a pre-weighed dish and dried in an oven at 180 °C until the weight of the dish no longer changes. The increase in weight of the dish represents the total dissolved solids and is reported in milligrams per liter (mg L^{-1}).

Toxin. (1) A microbial substance able to induce host damage. (2) Term applied to poisons in living systems.

Trace elements (nutrients). *See micronutrient.*

Trace gases. These are gases in the atmosphere, other than nitrogen and oxygen, that do not occur in large quantities but are significant to life on Earth or are important constituents of the chemical cycles in the atmosphere.

Translocation. In plants, the long-distance transport of water, minerals, or food; most often used to refer to food transport.

Transpiration. (1) The process by which water vapor is released from plants to the atmosphere primarily through the leaf *stomata*. (2) The loss of water molecules from the leaves of a plant; creates an osmotic gradient; producing tension that pulls water upward from the roots. [also see *evapotranspiration;* and *stoma*]

Triazone. A fertilizer that is a water-soluble compound of formula $C_3H_7N_3O$ which contains at least 41% total nitrogen.

Triple Superphosphate ($Ca(H_2PO_4)_2$) (0-46-0). A product that contains 40 to 50% available phosphoric acid. Triple superphosphate differs from ordinary superphosphate in that it contains very little calcium sulfate. (Also called treble superphosphate, concentrated superphosphate, double superphosphate, and multiple superphosphate). [also see *superphosphate;* and *monocalcium phosphate*]

Trommel. A revolving cylindrical sieve used for screening or sizing compost and mulch.

Trophic levels. Levels of the food chain. The first trophic level includes photosynthesizers that get energy from the sun. Organisms that eat photosynthesizers make up the second trophic level. Third trophic level organisms eat those in the second level, etc. It is a simplified way of thinking in relation to the food chain. Some organisms eat members of several trophic levels.

Tropical. A region in which the climate undergoes little seasonal change in either temperature or rainfall. Tropical regions of the Earth lie primarily between 30 degrees north and south of the equator.

Tropopause. The area where the temperature in the troposphere no longer decreases, indicating the boundary between the troposphere and the stratosphere.

Troposphere (Greek: tropos = turning, spharia = sphere). The lowest layer of the Earth's atmosphere ranging from the surface to the base of the stratosphere, with an altitude of 10 to 15 kilometers depending on the latitude. This is where all weather occurs. [also see *atmosphere; stratosphere;* and *mesosphere*]

Tuber. An enlarged, short, fleshy, underground stem. Examples are the Irish potato, cassava, and sweet potato.

Turgid. Swollen, distended; referring to a cell that is firm due to water and nutrient uptake.

Turgor pressure. The pressure within the cell resulting from the movement of water into the cell.

UAN (UAN-32) (32-0-0). A solution of urea and ammonium nitrate in water used as a fertilizer. The combination of urea and ammonium nitrate has an extremely low critical relative humidity (18% at 30°C) and can therefore only be used in liquid fertilizers. The most commonly used grade of these fertilizer solutions is UAN 32-0-0 (32%N) also known as UN32 or UN-32, which consists of 45% ammonium nitrate, 35% urea and only 20% water.

Ulmic Acid. The acid radical found in humic matter which is soluble in alkali. Ulmic acid is soluble in methyl ethyl ketone but is insoluble in water, any acid solution and methyl alcohol. Ulmates are the salts of ulmic acid. [also see *humates*; *humic acid*; *fulvic acid*; *humin*; *humus*; and *organic matter, soil*]

Unicellular. Single-celled organisms.

Unit. Twenty (20) pounds of plant food or one percent (1%) of a ton (AAPFCO). Represents 1% or 20 pounds of a nutrient in a 2000 pound ton of fertilizer. Example: a 10-20-15 fertilizer contains 10 units of N, 20 units of P_2O_5, and 15 units of K_2O per ton.

Unsaturated flow. The movement of water in a soil that is not filled to capacity with water. Water moves because of water-potential differences toward areas of lower water potential (drier soils).

Unsaturated solution. A solution that is not at equilibrium with respect to a given dissolved substance, and in which more of the substance can still dissolve.

Urea. ($CO(NH_2)_2$). $H_2N-\underset{\underset{O}{\parallel}}{C}-NH_2$ A fertilizer that is the commercial synthetic acid amide of carbonic acid; it should not contain less than 45% nitrogen. <u>Urea, the most concentrated dry nitrogen fertilizer</u>, is a non-acid-forming fertilizer which can be used with all types of soils and for any crop as a major nitrogen fertilizer; and in solution for additional foliar/irrigation fertilization. Urea is generally more effective than any nitrate forms of nitrogen for fertigation due to limited soil leaching. [also see *acid-forming fertilizer;* and *biuret*]

Urease. An enzyme that catalyzes the hydrolysis of urea into carbon dioxide and ammonia. The reaction occurs as follows:

$$CO(NH_2)_2 + H_2O \xrightarrow{\text{urease}} CO_2 + 2NH_3 \text{ (ammonia)}$$

Urease inhibitor. An additive to urea fertilizers which slows the rate at which urea hydrolyzes in the soil. A substance which inhibits hydrolytic action on urea by urease enzymes. When a urease inhibitor is applied to the soil the result is less nitrogen being lost by ammonia volatilization. [also see *ammonia volatilization; ammonification;* and *hydrolysis*]

Ureide. (1) Any of the derivatives of urea. (2) Any of several compounds derived from urea by the replacement of one or more hydrogen atoms by an acid radical.

Vadose zone. (1) The unsaturated soil between the soil surface and the underlying water table. The rooting depth is often called the root zone. (2) The unsaturated zone in a soil where chemical processes are at their most active. Its extent is determined partly by the content of the soil water, but it cannot extend beyond the water table, below which voids are completely filled with water.

Valence. (1) The number of hydrogen atoms which one atom of an element will combine with or displace. (2) The combining capacity of an atom or atoms. The total valence capacity varies from -3 to +6 in the active chemicals that are commonly found or used.

One significant result of valence is that only certain set combinations of atoms are possible. For instance, when nitrogen has a valence of -3, it can only combine with 3 hydrogens to form NH_3. There are no molecules that can be designated as NH, NH_2 or NH_4, etc.

Vapor. A substance in the gas state that is below its critical temperature but still suspended in air. It is possible for a vapor to be liquefied by increased pressure. Vapors are usually not visible.

Vascular. Consisting of, or containing vessels adapted for transmission or circulation of fluid.

Vascular plants. Vascular plants (also known as tracheophytes, or higher plants) are those plants that have lignified tissues for conducting water, minerals, and photosynthetic products throughout the plant. Vascular plants include the ferns, club mosses, flowering plants, conifers and other gymnosperms. Scientific names for the group include *Tracheophyta* and *Tracheobionta*, but neither name is very widely used.

Vascular tissue (plants). Xylem and/or phloem. [also see *xylem;* and *phloem*]

Vermicomposting. The process whereby worms feed on slowly decomposing organic materials (e.g., vegetable scraps) in a controlled environment to produce a nutrient rich soil amendment.

Vesicular arbuscular mycorrhiza. A common endomycorrhizal association produced by phycomycetous fungi of the family *Endogonaceae*. Host range includes most agricultural and horticultural crops and plants.

Virgin soil. A soil for all intents and purposes undisturbed from its natural environment; never cultivated.

Virulence. The degree of pathogenicity (the ability of a pathogen to produce an infectious disease in an organism) of a parasite.

Virus (Latin = poisonous liquid). Any of a large group of submicroscopic infective agents that typically contain a protein coat surrounding a nucleic acid core (RNA or DNA) and are capable of growth only in a living cell.

Volatilization. The evaporation or changing of a substance from liquid to vapor. [also see *ammonia volatilization*]

Volcanic ash. See *ash, volcanic*.

Waste, agricultural. Waste materials produced from the growing of plants and raising of animals; including manures, bedding, plant stalks, hulls, leaves, and vegetable matter.

Water. Water consists of two hydrogen atoms and one oxygen atom (H_2O). Water covers ~70.8% of the Earth's surface, 60% to 70% of the world's weight, and regenerates and redistributes through evaporation and other atmospheric processes. Water is involved in electrical charge separation because it has two types and positions of atoms giving it a net dipole movement. Water vapor also absorbs 17% of solar radiation in the troposphere, thus making it one of the two principal greenhouse gases.

Of the solar energy absorbed by the Earth's surface, a little more than half goes into latent heat, which is heat absorbed by water due to its transformation from a liquid to a gas. When these molecules condense back into a liquid, usually higher in the atmosphere, they released that energy back into the atmosphere as local warming.

Water cycle. The process by which water is transpired and evaporated from the land and water, condensed in the clouds, and precipitated out onto the Earth once again to replenish the water in the bodies of water on the Earth. [also see *hydrologic cycle*]

Approximate Amounts of Water Held by Various Soils		
Soil Texture	Inches of Water Held per Foot of Soil (inches)	Maximum Rate of Irrigation per Hour of Bare Soil
Sand	0.5-0.7	0.75
Find Sand	0.7-0.9	0.60
Loamy Sand	0.7-1.1	0.50
Loamy Fine Sand	0.8-1.2	0.45
Sandy Loam	0.8-1.4	0.40
Loam	1.0-1.8	0.35
Silty Loam	1.2-1.8	0.30
Clay Loam	1.3-2.1	0.25
Silty Clay	1.4-2.5	0.20
Clay	1.4-2.4	0.15

Western Fertilizer Handbook. 9th ed. 2002.

Water, gravitational. Water that moves into, through, or out of a soil under the influence of gravity.

Water potential, soil. The amount of work (usually given in kilopascals) an infinitesimal quantity of soil water can do in moving from the soil to a pool or pure, free water at the same location and at normal atmospheric pressure. Soil water potential is mostly matric potential. The total water potential consists of the following potentials:

Matric potential. That portion of the total soil water potential due to the attractive forces between water and soil solids as represented through adsorption and capillarity. It will always be negative.

Osmotic potential. That portion of the total soil water potential due to the presence of solutes in soil water. It will generally be negative.

Gravitational potential. That portion of the total soil water potential due to differences in elevation of the reference pool of pure water and that of the soil water. Since the soil water elevation is usually chosen to be higher than that of the reference pool, the gravitational potential is usually positive.

Water soluble potash (K_2O). On a fertilizer label with N, P, & K, the K stands for the percent of water soluble potash K_2O in the fertilizer. [also see *available phosphate (P_2O_5)*]

Water table. (1) Upper level of groundwater (water collected underground in porous rocks). Water that is above the water table will drain downwards; a spring forms where the water table meets the surface of the ground. The water table rises and falls in response to precipitation and the rate at which water is extracted, for example for irrigation and industry. (2) That level below which the soil material is saturated with water. [also see *perched water table*]

Water use efficiency. A measure of the crop production per unit of water used, irrespective of water source, expressed in units of weight per unit of water depth per unit area. The concept of utilization applies to both dryland and irrigated agriculture.

Also, crop irrigation accounts for 70% of the world's fresh water use. Agriculture in most countries is extremely important both economically and politically, and water subsidies are common. However, conservation advocates have urged removal of all subsidies to force farmers to grow more water-efficient crops and adopt less wasteful irrigation techniques. It has become imperative for farmers to grow more water-efficient crops and adopt less wasteful irrigation techniques. Optimal water efficiency means minimizing losses due to evaporation, runoff or subsurface drainage.

Water vapor. Water present in the atmosphere in gaseous form; the source of all forms of condensation and precipitation. Water vapor, clouds, and carbon dioxide are the main atmospheric components in the exchange of terrestrial radiation in the troposphere, serving as a regulator of planetary temperatures via the greenhouse effect. Approximately 50% of the atmosphere's moisture lies within about 1.84 kilometers of the Earth's surface, and only a minute fraction of the total occurs above the tropopause. [also see *tropopause*]

Waterlogged. Saturated with water, usually developing anaerobic conditions. Most or all pore space is filled with water, natural or induced by humans. [also see *anaerobic*]

Watershed. The total land area that contributes water to a river, stream, lake, or other body of water. Sometimes synonymous with drainage area, drainage basin, and catchment.

Weathering. All physical disintegration, chemical decomposition, and biologically induced changes in rocks or other deposits at or near the Earth's surface by atmospheric or biologic agents, or circulating surface waters with essentially no transport of the altered material. These changes result in disintegration and decomposition of the material.

Wetlands. A transitional area between aquatic and terrestrial ecosystems. Soil areas that have evidence of saturated conditions part of the year (ponded water and wet-area plants (e.g., cattails).

Wilting point. See *permanent wilting point (percentage)*.

Wood-overs. Also called "compost-overs." Wood-overs are the large, woody parts of the compost pile that have not completely broken down and will not pass through a minus one-half inch Trommel screen. [also see *Trommel*]

Xenobiotic. A compound foreign to biological systems. Often refers to human-made compounds that are resistant or recalcitrant to biodegradation and decomposition.

Xerophytes. Plants that grow in extremely dry areas.

Xeroscaping. The practice of landscaping with slow-growing, drought-tolerant plants.

Xylem (plural xylems). In a plant, the main water conducting tissue from the roots upward, consisting of lignified tracheids or vessels, and which also acts as a supporting tissue. Secondary xylem = wood. [also see *phloem* and *vascular*]

Yeast. A fungus whose thallus consists of single cells that multiplies by budding or fission. [also see *mold*]

Yellow-green algae. See *algae, yellow-green*.

Zeolites. Microporous silicate or aluminosilicate structured minerals that can act as an absorbing filter or sieve on a molecular level. Mainly used by the petroleum industry for the cracking of petroleum or use as a filter against various compounds.

Zinc (Zn). A metallic micronutrient present in the soil and absorbed by plants as the Zn^{2+} ion. Zinc's oxidation state in the soil remains to same. One of the first micronutrients recognized as essential, zinc aids synthesis of plant growth substances and enzyme systems, and is essential for promoting certain metabolic reactions. It is also necessary for production of chlorophyll and carbohydrates.

The metal is not translocated within the plant, so deficiency symptoms appear first on the younger leaves and other young plant parts. Zinc becomes less available as soil pH increases, and high soil phosphorus availability also increases zinc deficiency. Much of the soil's available zinc is associated with the organic fraction and deficiencies may be associated with low soil organic matter. Cool soil temperatures in early spring can intensify the need for zinc. When soils are cold, organic matter does not decompose and Zn^{2+} is not released and available for crop growth. [also see *plant nutrient, essential;* and *cation*]

Zinc nitrate ($ZnNO_3)_2 \cdot 6H_2O$). A highly soluble fertilizer compound used in solution forms as a source of zinc, especially in liquid fertilizers. All nitrates are soluble including zinc nitrate. The solubility of $(ZnNO_3)_2 \cdot 6H_2O$ is ~1850g per liter (~15.4 lbs gal^{-1}) meaning it is very soluble in water.

Zinc sulfate ($ZnSO_4 \cdot 7H_2O$). The zinc salt of sulfuric acid. A solid water-soluble fertilizer material, containing 36% zinc, used as a source of zinc for plants. Zinc sulfate is 99.9% water-soluble (~8.05 lbs gal^{-1}).

Annotated Bibliography

AAPFCO: Association of American Plant Food Control Officials.

Brady, N. C., and R. R. Weil. 2008. The Nature and Properties of Soils. 14th Edition. Pearson-Prentice Hall, Upper Saddle River, New Jersey.

Brewer, R. 1976. Fabric and Mineral Analysis of Soils. John Wiley & Sons, New York.

Bohn, H., B. McNeal, and G. O'Connor. 2001. Soil Chemistry. 2nd Ed. John Wiley & Sons, New York.

CRC Handbook of Chemistry and Physics. 2010. 91st Edition. CRC Press, Boca Raton, Florida.

Keys to Soil Taxonomy. 2010. 11th Edition. Natural Resources Conservation Service, United States Department of Agriculture.

Lorenz, O. A., and D. N. Maynard. 1988. Knott's Handbook for Vegetable Growers. 3rd Edition. John Wiley & Sons. New York.

Miller, R. W., and D. T. Gardiner. 2007. Soils in Our Environment, 11th Edition. Prentice Hall, New Jersey.

Rosen, C. J., and P. M. Bierman. 2005. Using Manure and Compost as Nutrient Sources for Fruit and Vegetable Crops. University of Minnesota Extension Bulletin. M1192.

Soil Survey Manual. USDA Handbook No. 18. 1993. Soil Conservation Service, United States Department of Agriculture.

Thorup, R. M. Orthro Agronomy Handbook. 1984. Chevron Chemical Company. San Francisco, California.

Western Fertilizer Handbook. 2002. 9th Edition. Prentice Hall, New Jersey.

www.ingramcontent.com/pod-product-compliance
Lightning Source LLC
Chambersburg PA
CBHW030934180526
45163CB00002B/567